Oxford Physics Series

General Editors
E.J. BURGE   D.J.E. INGRAM   J.A.D. MATTHEW

Oxford Physics Series

G.A. JONES

University Lecturer in Nuclear Physics,
and Fellow of Wolfson College, Oxford

The Properties of Nuclei

Clarendon Press · Oxford
1977

Oxford University Press, Walton Street, Oxford OX2 6DP

OXFORD LONDON GLASGOW NEW YORK
TORONTO MELBOURNE WELLINGTON CAPE TOWN
IBADAN NAIROBI DAR ES SALAAM LUSAKA ADDIS ABABA
KUALA LUMPUR SINGAPORE JAKARTA HONG KONG TOKYO
DELHI BOMBAY CALCUTTA MADRAS KARACHI

ISBN 0 19 851828 5

© Oxford University Press 1977

Printed in Great Britain by Thomson Litho Ltd., East Kilbride, Scotland.

Editors' foreword

This book fits closely into the network of topics of the Oxford Physics Series and at the same time preserves an integrity of its own. The earlier complementary core texts include *Radiation and quantum physics*, and *Atomic nuclei and their particles* with supporting texts on *Electromagnetism* and *Special relativity*, and *Atoms and their structure*.

The volume *Basic quantum mechanics* will provide the necessary introduction to many sections of this new text on *The properties of nuclei*, the level of which is pitched to take over where the corresponding sections of *Atomic nuclei and their particles* leaves off.

Dr. Jones concentrates his attention on the available data and their interpretation, developing a natural sequence from the properties of nuclear forces, through nuclear models to nuclear decay and reactions including fission and fusion. The study of these topics is undertaken in second- or third-year courses in physics, applied physics, and nuclear engineering and challenges the student to appreciate both the power of the available experimental and theoretical methods and also the wide range of opportunities for research in the nuclear many-body problem. The sureness of approach found in this treatment illuminates the scene in a rewarding fashion which reflects the author's experience of teaching and developing the subject over a number of years.

March 1977.   E.J.B.

Preface

The level I have aimed at in this book is that of the nuclear-structure course given to Oxford students in the final term of their second year. In their third year some, but not all, pursue a more advanced course, according to their choice of options. The book therefore reflects the requirements of 'mainstream' undergraduate physics courses. It presupposes a certain familiarity with particle accelerators and radiation detection.

The content is chiefly devoted to the static and quasi-static properties of nuclei and nuclear excited states. But I firmly believe that all physics students should have some notion of how a nuclear reactor works, and so a chapter on nuclear reactions has been included in addition to a brief account of fission (and fusion).

In a book of this length much has to be taken on trust; quantum-mechanical formulae have been used freely with no more than a passing reference (the particular reference is to a text which I found useful as a student and still do as a teacher, but a more recent publication may appeal to the reader). Whatever the choice, I hope my book will encourage the student to acquire a working knowledge of basic quantum mechanics, which forms the framework of his (her) studies.

I am indebted to members of the Department of Nuclear Physics, Oxford, who read and commented on the script, and especially to Dr. M.G. Bowler with whom I had many valuable discussions. I should also like to thank one of the editors of this series (E.J.B.), who,

I feel, put more time than his editorial duties deman-
ded into a critical appraisal which I much appreciated.

Finally, a number of the figures were taken from,
or based on, published work. The figure captions in-
dicate, where relevant, the author and the work. My
thanks are also due to the publishers for permission
to reproduce published work: They are: Fig. 1.2,
Springer Verlag; Figs. 2.1, 3.8, 3.9, 4.3(a),(b),
North Holland; Figs. 2.4, 6.3, American Institute of
Physics; Fig. 3.2, Pergamon Press; Fig. 3.3, Annual
Reviews; Fig. 3.4, Academy Munksgaard; Fig. 4.2,
McGraw Hill; Fig. 5.2, John Wiley and Sons; Fig. 6.4,
Brookhaven National Laboratory; Fig. 7.3, W.A. Ben-
jamin.

G.A.J.

Contents

CONTENTS

1. General properties of nuclei

INTRODUCTION

From early measurements on Coulomb scattering it is known that the atomic nucleus is a small central particle (radius $\ll 10^{-11}$ cm) which possesses virtually the whole mass of the atom and which carries a positive charge equal to the atomic number of the atom. From later work it is known that the nucleus is built up of protons and neutrons; the former supply the positive charge, whilst both contribute to the mass. From general quantum-mechanical considerations such a confined system will possess energy levels characterized by quantum numbers. Under normal circumstances, the nucleus will be in its ground state, and this chapter is concerned with nuclear properties in this state. To determine the properties of the nucleus in other (excited) states it is necessary to set up such states by means of nuclear reactions.

Thus a particular nucleus (in its ground state) must be described in terms of the quantum numbers 'angular momentum' (also loosely called spin) and 'parity' (to be explained later in the chapter). This is the minimum number of quantum numbers required by the symmetry properties of the system. There are others, but they are approximate or model-dependent, i.e. they depend on a simplified picture of the nucleus, and, unlike the corresponding atomic system, no one model stands out sufficiently to attach extra significance to its quantum numbers.

In addition, a system of finite extent can be described in terms of moments of the matter distribution. In the case of the nucleus these moments are usually determined from its interaction with the atomic electrons which is electromagnetic in origin, so the moments measured are electric and magnetic multipole moments. A system can have an electric dipole moment only if it has an up-down directional asymmetry in space, e.g. the molecule HCl, but not Cl_2, can have a permanent electric dipole moment. For nuclei no such asymmetry is believed possible, so nuclei do not possess such a moment. Since the components of the nucleus, protons and neutrons, have intrinsic spins and magnetic dipole moments, the nucleus itself can also have a magnetic moment. The two important moments of the nucleus, apart from the electric monopole, are therefore its magnetic dipole and electric quadrupole moments.

Whilst nuclear physics deals with all nuclei, of especial interest are the so-called 'stable nuclei' These are the nuclei which exist in the world around us, and this is quite specific, though 'stable' itself is much less definite. A nucleus may be unstable against emission of a neutron or a proton, or a composite particle such as an α-particle, or against β-decay. Of these neutron emission is very fast ($\tau \sim 10^{-16}$ or less), proton emission generally less so but still very fast, whilst the latter two can have lifetimes varying over a vast range depending on the energy available ($\tau \sim \mu s$ to $\sim 10^{20}$ years). By 'stable' a lifetime of the order of the age of the universe ($\sim 10^{10}$ years) is implied; sometimes it has the restricted meaning 'stable against the β-decay process'.
THE SIZE

The original measurements made by Rutherford and

his co-workers have now been augmented by more accurate
work, which has not only measured an effective radius
for the nucleus but has also obtained a measure of the
rate of fall-off of density, or charge, at the periphery. The measurements fall into two classes: (1)
those that determine the charge distribution and (2)
those that determine the matter distribution.

The charge distribution

These measurements are made with charged leptons
as probes. Leptons form a class of fundamental particles which interact only weakly with protons and
neutrons, except for the long-range Coulomb term where
relevant.[†] Thus the interaction between an electron
or a muon and nuclear matter is completely dominated
by the Coulomb interaction. The greatest source of
information is the elastic scattering of electrons.
Accurate measurements of optical and X-ray spectra
and muonic X-ray spectra have also given information on
the nuclear size.

Electron elastic scattering

Outside the nucleus of charge Ze the electron experiences a force $\propto Z/r^2$, but inside the nucleus
(assumed spherical) the force is $\propto Z'(r)/r^2$, where
$Z'(r)$ is the charge contained in the sphere of radius
r. Thus small-angle scattering, which corresponds to

[†]Interactions between fundamental particles can be
put into the categories strong, electromagnetic, weak
and gravitational, which are roughly in strength as
$10 : 1/137 : 10^{-7} : 10^{-39}$ (see e.g. Perkins 1972).

a classical trajectory rather distant from the nucleus, will obey the Rutherford formula and therefore give no information about the distribution of charge, whilst large-angle scattering will show a significant deviation from the formula. The ratio to Rutherford scattering as a function of angle is determined by the form of $Z'(r)$ and therefore by the charge distribution. From the uncertainty principle, if we wish to attach significance to changes of density over a distance Δr, then the incident particle must be capable of transferring momentum $\hbar q$ to the target nucleus such that $q\Delta r \sim 1$; obviously the wavenumber of the incident beam must exceed q. For $\Delta r \sim 10^{-13}$cm, electrons of kinetic energy > 200 MeV are required. Some twenty years ago, measurements were made at 50 MeV and were capable of defining a suitably averaged charge radius only; more recently, the energy range has been extended to 600 - 1000 MeV and the radial dependence of the nuclear charge distribution has been revealed in fine detail.

X-ray and optical spectra

Since the nucleus has finite extent its field of force on the atomic electrons will not be $\propto 1/r^2$ when the electron is within the nucleus. If therefore the atomic energy levels can be calculated accurately for a point nucleus and compared with their measured values, the nuclear size can be deduced. But the calculation is an unsolved many-body problem. For X-ray transitions the screening effect of the outer electrons is not large and can be approximated to; but for the tightly-bound inner electrons there are uncertainties even in the solution of the two-body problem, whilst for the valence electrons, although these latter un-

certainties do not arise, the screening effects cannot
be calculated to sufficient accuracy. In certain cases
useful information can be derived. It is reasonable
to assume that, for a comparison between spectral lines
from isotopes of the same element, the differences will
arise from the differences in nuclear mass and size
only. For light elements the mass difference produces
effects large enough to measure, but for heavier elements the measured effects must be ascribed to differences in nuclear charge distribution. Thus the
method of isotope shift gives information on the way
the charge distribution changes on adding neutrons
to a nucleus.

Muonic X-rays
    Negative muons μ^-, like electrons, can be captured
into atoms and interact electromagnetically with nuclei.
Since they have a mass $\sim$ 200 m_e (where m_e is the mass
of the electron), their Bohr orbits will have 1/200th
of the radius of the equivalent electron orbits. For
a heavy element such as lead, the lowest Bohr orbit
(since no orbits are occupied by other muons, the
exclusion principle does not operate here) is such
that the muon spends most of its time within the nucleus. The size effect is therefore large, reducing the
lowest p → s transition from $\sim$ 16 MeV for a point
nucleus to $\sim$ 6 MeV as measured. The muon is, of course,
unstable and decays spontaneously to an electron and
two neutrinos (ν) with a mean life of $\sim$ 2 μs. In the
lowest Bohr orbit in lead it decays in about half this
time because of the competing mode $\mu^- + p \rightarrow n + \nu_\mu$,
but during its lifetime it traverses several metres of
solid nuclear matter (density $\sim$ 3 × 10^{17} kg m^{-3}), thus

emphasizing the weakness of the weak interaction.

From a theoretical aspect, the computational difficulties are the same as those encountered in inter-
preting electronic X-rays, but the uncertainties are
reduced. Obviously the screening effects of the
atomic electrons will only be of the same order as for
the K-electrons, whereas the nuclear interaction energy is increased by the factor 200. It also turns out
that the uncertainties in the two-body problem are reduced by the same factor. Since the bound muons have
effectively a long wavelength (∿ size of the Bohr
orbit) the measurements lead to an average nuclear
radius and give no information on the charge distribution. But the average they give is very precise,
and nowadays it is used as a datum in the analysis of
electron scattering, acting as a constraint when look-
ing for the finer details of the charge distribution.

The matter distribution

If the nuclear probe interacts strongly with protons and neutrons then the size determined from the
interaction will be that of the nuclear matter as a
whole. We have also the added complication that the
strong interaction has a range which, though short,
is not short compared to the nuclear size, and due
allowance must be made for this interaction when deducing the mass distribution.

The historical example of this kind of measurement is the elastic scattering of α-particles, which
led to the first determinations of the nuclear size.
Other charged particles have since been used, e.g.

protons, carbon and oxygen nuclei (in the form of partially stripped ions), and other heavy ions. From a
theoretical point of view, the simplest process is the
scattering of neutrons, but experimentally this is
rather difficult, since it involves three nuclear interactions - one to produce the neutron, another to scatter it, and a third to detect it. Other methods involving the electrostatic term are of historical importance: they are the lifetime for α-radioactivity
and the determination of the electrostatic energy of
the nucleus from the energy difference between mirror
pairs. Recently measurements of K-mesic X-rays have
proved to be a promising method of probing the nuclear
surface.[†]

Scattering of charged particles

Experimentally, the scattering cross-section (see
Appendix A) is measured as a function of angle for a
fixed beam energy or as a function of energy at a fixed
angle. In either case the ratio of measured cross-
section to Rutherford cross-section is plotted as a
function of the variable. The ratio will be unity over
a range of angles from near zero up to a certain angle,
beyond which it drops off, indicating absorption of
the incident particle. Alternatively, for constant
angle the ratio will be unity up to a certain energy,
beyond which it drops off. However, for protons or
even α-particles on light nuclei it is possible for
the cross-section to rise above the Rutherford value,

[†]K-mesons, which can be positively or negatively
charged or neutral, have a mass $\sim 1000\ m_e$ and interact strongly with nucleons.

because such particles stand a good chance of being emitted after absorption. For heavier ions, or protons and α-particles on heavy nuclei, emergence of the absorbed particle is improbable.

The region of fall-off is correlated with the classical trajectory passing through the periphery of the nucleus, immediately giving an idea of the size of the nucleus. How to obtain a more precise determination of the size was uncertain before the development of the optical model (see Chapter 3). We should notice that, in the interpretation of the results, in addition to making allowance for the range of nuclear forces it is also necessary to take into account the finite size of the incident particle.

Neutron scattering

When fairly energetic neutrons ($\sim$ 10 - 30 MeV) impinge on nuclei they are strongly absorbed from the initial beam, though they may subsequently be emitted from the resulting excited nuclei. The degree to which they are emitted depends upon the number of competing modes of decay (see Chapter 6). If there are many such modes, and there are for $E \sim$ 10 - 30 MeV, elastic scattering will be reduced to a minimum. Elastic scattering cannot be zero since an extension of Babinet's theorem in optics to this nuclear wave problem indicates that absorption without scattering is impossible. This scattering is predominantly through small angles (up to $1/kR$ as in optical diffraction). The absorption cross-section can be conveniently measured by a beam-attenuation technique. A neutron detector placed in the path of a collimated neutron beam determines the beam flux before and after interposition of the absorbing target. If the angle subtended by the detector at the target is greater than the angle of diffraction, then diffracted neutrons are detected, and the beam attenuation is a true measure of nuclear absorption. A simple theory gives $\sigma_{abs} = \pi(R + \lambdabar)^2$, where the additional term $\lambdabar$, the wavelength of the projectile, expresses the uncertainty with which we can locate a neutron. A more sophisticated analysis can be carried out in terms of the optical model.

α-Radioactivity

From consideration of the mass equation (see pp. 12-20), all nuclei above $A \sim 150$ are unstable, in principle, and can break up into two smaller units. Because of its high intrinsic binding, the α-particle is usually one of these units. The process involved will be discussed in more detail in Chapter 4, but for the present it can be summarized by stating that the probability of α-particle emission depends on two factors: the intrinsic nuclear factor and the barrier penetrability. The latter is a rapidly varying function of the nuclear radius, which can therefore be determined if the former is known - the measured lifetime is known with great precision. Unfortunately, the intrinsic factor is a notional one and cannot be measured; theoretical estimates are uncertain, with the result that the nuclear radius deduced is qualitative only. It is probably more meaningful to use the data, and a nuclear radius from other data, to deduce the intrinsic nuclear factor.

Electrostatic energy differences of mirror nuclei

Mirror nuclei are nuclei in which the proton number and neutron number are interchanged, e.g. ^{13}C and ^{13}N having $Z = 6$, $N = 7$ and $Z = 7$, $N = 6$. Most of the measurements refer to mirror pairs having $|N - Z| = 1$.

It is believed (Chapter 2, p.42) that the specifically nuclear interactions will be the same for such pairs, so their mass difference will arise from the electrostatic interaction and the n-p mass difference. The former term is discussed in greater detail in the next section, for the moment a uniform distribution of charge in the nucleus is assumed to give for the mass difference

$$\Delta M = M(A,Z+1) - M(A,Z) = m_p + m_e - m_n + \frac{3}{5} \frac{1}{4\pi\varepsilon_0} \frac{(2Z+1)e^2}{R},$$

where the M and m are defined in the next section.

The mass difference can be measured accurately from the β^+-decay spectrum of $(A, Z+1)$, since

$$\Delta M = m_e + E_{kin}$$

where E_{kin} is the maximum kinetic energy available to the β^+; or from the threshold energy for neutron production by protons on (A,Z),

$$M(A,Z) + m_p + m_e = M(A,Z+1) + m_n + Q$$

$$\Delta M = m_p + m_e - m_n - Q$$

Thus the Q-value of the (p,n) process is the negative of the electrostatic energy difference.

From the data, $R = R_0 A^{1/3}$ is found to hold good, but the value deduced for R_0, namely 1·46 fm is rather large compared with deductions from other processes (see the end of this section). This reflects the dubious supposition that the nuclear charge is uniformly distributed within the nucleus. Since the data on mirror nuclei refer only to light nuclei ($A < 40$) surface effects will be expected to be large.

K-mesic X-rays

From the discussion of muonic X-rays, it follows that the K^--mesic Bohr orbits will have an even smaller radius. In addition, the K^--meson will interact strongly with the nucleus, so nuclear absorption will compete with X-ray emission. Absorption will occur from highly excited Bohr orbits when the K-meson overlaps the tail of the matter distribution. These measurements therefore focus attention on the details of the matter distribution at the surface of the nucleus.

Conclusions on size

Analyses of the earlier data, which were capable only of defining an effective nuclear radius, gave the result that the radius is proportional to the cube root of the mass number A, i.e. that the nuclear density is a constant independent of the nuclear size. Later measurements indicate that, although this is true for the central region, near the periphery the density falls off smoothly and quickly to zero. The precise nature of the fall-off is not determined from the data, but it has proved convenient to represent it by the (Fermi) function

$$(\rho)r \propto [1 + \exp \{\frac{(r-R)}{a}\}]^{-1},$$

where R, the radius at half-density, is referred to as

the nuclear radius and a is a surface thickness parameter which defines the region over which the density
is falling off. This latter parameter appears to be
independent of nuclear size, whilst $R \propto A^{1/3}$.

The measurements referring to the charge distribution are of greater precision than those referring
to the matter distribution, but both agree on the
values of R and a; there appears to be no significant
difference between the distributions of neutrons and
protons within the nucleus. The data may be summed
up by giving the charge radius as $R = R_0 A^{1/3}$ where
$R_0 = 1 \cdot 1$ fm and $a \sim 0 \cdot 5$ fm. The latter figure is
not known accurately. The matter distribution will
have the same parameters, but both are less accurate.

An objection may be raised, by the more dis-
cerning student after reading on, that nuclei have
been treated as spherical whereas some of the more
massive nuclei have quite large quadrupole moments.
The reason is that the nuclei have not been aligned
in their targets during the experiments described,
and so the results give a time- and space-averaged picture of the nucleus which will have spherical symmetry even when it is intrinsically non-spherical. The
averaging process may well be expected to produce a
wider region of fall-off i.e. a greater value for a.
There appears to be some indication of this.

THE MASS AND STABILITY

Historically, the masses of neutral atoms were
determined relative to (atomic) $^{16}O \equiv 16$ atomic mass
units and later relative to (atomic) $^{12}C \equiv 12$ U (uni-
fied atomic mass unit), by chemists using chemical
reactions. The measurements were later improved upon

using physical techniques involving the trajectories of ionized atoms through electric and magnetic fields. With this information available the nuclear physicist found it expedient to use atomic masses in his equations for nuclear reactions, since conservation of charge ensures that the same number of electrons will appear on each side of an equation (though not for β^+-decay, see p. 21). Balancing the electron number, however, does not allow completely for the difference between atomic and nuclear masses, since no account has been taken of the binding energy of the atomic electrons - which, though small, can add up to $\sim 0 \cdot 5$ MeV for uranium. If this appeared directly as an error in the determination of Q-values in nuclear reactions then it would be serious. But most reactions are concerned only with small changes of A and Z and therefore only with small changes in this term, which, if neglected, could introduce errors of order 20 keV for a-decay in the region of uranium when using nuclear masses - though the correction is well known and accurate to a fraction of 1 keV. On the other hand, use of atomic masses largely eliminates such effects since the mass measurements are done on ions which are ionized in the valency shell and thus already include all the large contributions to electronic binding. For this reason the atomic scale is to be preferred. However, the student may come across the usage of nuclear mass (Oxford University Examination Papers are a case in point), and so, although the atomic scale will be employed here, comparisons will be made between the two scales. The nomenclature employed will be that M will refer to atomic masses and m will refer to nuclear masses. Since the masses quoted for the proton and neutron are usually m_p and m_n, rather than M_H ($M_n = m_n$ anyway), the mass equations will contain $Z(m_p + m_e)$ rather than ZM_H.

From the equivalence of mass and energy, the mass of a nucleus consisting of Z protons and N neutrons ($Z + N = A$) will be less than the sum of the masses of the individual nucleons by the energy of binding (expressed in u ($1 \cdot 66 \times 10^{-27}$ kg) - it is in fact more convenient to re-express the masses in MeV and since, in u, the mass values are close to the integers A, it is usual to give the mass excess in MeV, i.e. $\{M(A,Z) - A\}$ u converted into MeV). Thus

$$M(A,Z) = Z(m_p + m_e) + (A-Z)m_n - B.$$

It is necessary therefore to consider all contributions to the binding energy. This is done in an empirical way, but guided by physical facts and theories, in the mass equation of von Weizsäcker.

From the fact that nuclear density (neglecting surface effects) is a constant a nucleus can be compared with a liquid drop, for which it is known that the binding energy (again neglecting surface effects which are usually small for liquid drops) is proportional to its volume, since all molecules deep in the liquid are in the same environment. Hence, by analogy, the leading term in the mass equation representing the binding energy will be $-\alpha A$, where α is some undefined constant. All constants introduced into the mass equation will be defined to be positive.

A second term in the binding energy also arises from the analogy with a liquid drop. The forces which keep the liquid drop together (van der Waals forces)

are short range, so that the volume binding energy
term is dominated by the interactions between near
neighbours. At the surface a molecule will have fewer
near neighbours than a molecule in the interior and
so will be less tightly bound, giving a surface-energy
term and a surface tension, the former being proportional to the surface area. It results in a lower
total binding of the drop and therefore appears as a
positive term in the mass of the drop. Since the surface area of a sphere $\propto r^2$, and, reverting to the
nucleus, $r \propto A^{1/3}$ the surface-energy term will appear
as $+\beta A^{2/3}$ in the nuclear mass equation. As a large
fraction ($\sim \frac{1}{2}$) of nucleons are near the surface this
will be expected to be a considerable fraction of the
volume term.

The analogy with a liquid drop neglects the charge
on the nucleus. On attempting to remedy this by char-
ging the drop, the analogy breaks down, since the
charge migrates to the surface of the drop whereas,
from the previous section, it is known that the charge
density of a nucleus is proportional to the matter
density. This illustrates the difficulties in making
classical models of quantum-mechanical systems; obviously the uncertainty principle operates to reduce
the energy when both protons and neutrons make full
use of the space available. Assuming that the nuclear
charge is uniformly distributed over the volume of a
sphere then its electrostatic potential energy can
easily be shown to be $\frac{3}{5} (Ze)^2/4\pi\varepsilon_0 R$. If the discrete
nature of the charge on the proton is taken into ac-
count, perhaps Z^2 should be replaced by $Z(Z-1)$. This
refinement is probably an improvement but, except for
small nuclei, it makes little difference in the mass

equation. In any case there are larger factors of ignorance: the charge distribution is known to vary with radius within the nucleus, and there could be a correlated motion which allows protons to avoid each other yet retains the required charge distribution. The electrostatic term is therefore introduced into the mass equation as $+\varepsilon Z^2 A^{-1/3}$, where the (unknown) constant ε is used to account for these uncertainties.

At this stage the mass equation is

$$M(A,Z) = Z(m_p + m_e) + (A-Z)m_n - \alpha A + \beta A^{2/3} + \varepsilon Z^2 A^{-1/3}.$$

Minimizing this function with respect to Z at constant A gives, for stable nuclei, the approximate result $Z = A^{1/3} (m_n - m_p - m_e)/2\varepsilon$. The important factor is $Z \propto A^{1/3}$, which is far too slow a variation with A (see Fig.1.1). $A = 32$ has $Z = 16$ stable, so the above

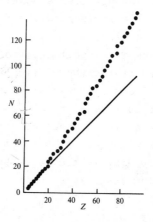

Fig.1.1. Plot of N against Z of most stable nucleus for given A. For convenience only every sixth A-value is plotted. The full line shows $N = Z$.

formula would give $Z = 32$ stable for $A = 256$, to be
compared with the naturally stable $^{238}_{92}$U having $A = 238$,
$Z = 92$.

It is interesting to note from Fig.1.1 that, up
to $A = 40$, the line of stability has $N = Z$. Above
this value N increases more rapidly than Z, presumably
because of the electrostatic term in the mass equation.
It would appear that a term $\propto (N-Z)^2$, i.e. $(A-2Z)^2$ is
also required, which if of positive sign would mitigate
in favour of $N = Z$ for stable nuclei. One need not
look far for the origin of such a term; the Pauli
principle will ensure that the nucleus (in the absence
of the electrostatic term) should have $N \sim Z$ in order
to utilize the lowest levels available to both species.
To proceed further it is necessary to invoke a model.
A convenient model, which incorporates the Pauli principle, is to consider the neutrons and protons as two
non-interacting degenerate Fermi gases inside the
nucleus. Neglecting the electrostatic term and the
n-p mass difference, which are already in the mass
equation, the model instantly leads to $N = Z$ for minimum energy. Expanding the energy function from the
point $N = Z$ gives a leading term $\propto (N-Z)^2/A$. Hence
a symmetry term $+\gamma(A-2Z)^2/A$ is added to the mass
equation.

An extra term is needed before attempting to use
the mass equation to determine which nuclei are stable.
In Appendix B it is shown, under approximation, that
the balance between terms so far developed in pre-
dicting the heavy stable nuclei is so finely poised
that the nearest nucleus (of given A) to the line of
stability will be most stable. Thus, starting from an
even-Z even-N stable nucleus, such that the next

stable mass is obtained by adding a neutron, then the
second nucleon added to form a stable nucleus will be
a proton more often than a neutron since, if (A,Z) is
on the line of stability then $(A + 2, Z + 1)$ will be
nearer that line than $(A + 2,Z)$. Thus, for any given
A, only one stable nucleus will be expected, and of
those with even A, stable nuclei will occur as often
with odd Z as with even Z.

Nuclear charts show that there are 166 even-Z
even-N stable nuclei but only 4 of odd Z, odd N, and
these are very light: $^{2}_{1}H$, $^{6}_{3}Li$, $^{10}_{5}B$, $^{14}_{7}N$. Conditions
for very light nuclei do not conform to those under
which the mass equation was developed, and in addition
the systems have few widely-spaced energy levels which
could affect the balance between Z and N. Of the 105
odd-A stable nuclei, 50 are found to have odd Z.

In order to account for this distribution of
stable nuclei, a pairing energy is postulated and is
accounted for in the mass equation by a small term
$\delta(A,Z)$, where δ is a function which has the value zero
for all odd A and takes a negative value when both A
and Z are even but a positive value when A is even and
Z is odd. Its magnitude drops off as A increases, and
this is to be expected since, on average, the two similar
pairing nucleons will be further apart for large A than
for small A.

The atomic mass equation therefore reads

$$M(Z,A) = Z(m_p + m_e) + (A-Z)m_n - \alpha A + \beta A^{2/3} +$$
$$+ \frac{\gamma(A-2Z)^2}{A} + \frac{\varepsilon Z^2}{A^{1/3}} + \delta(A,Z) \quad . \tag{1.1}$$

This five-parameter formula has been fitted to the

available mass data (several hundred nuclei neglecting the very light ones, $A < 20$) to give values:[†]

Parameter	α	β	γ	ε	δ
Value (MeV)	15·6	17·2·	23·3	0·7	$33\frac{1}{2} A^{-\frac{3}{4}}$

The fit is shown in Fig.1.2, where it was more convenient to present the binding energy per nucleon (B/A)

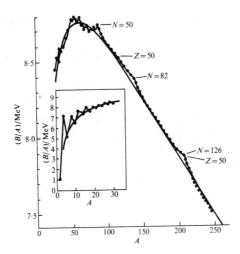

Fig.1.2. Plot showing the binding energy per nucleon as a function of A. Locations of magic numbers are indicated. The inset continues the plot and illustrates α-particle structure in light nuclei. (Based on *Wapstra* (1958). Atomic masses of nuclides, *Handb.Phys*. XXXVIII/I.

[†]The values of the constants vary from source to source. Usually no errors are given so the significance of the variations cannot be assessed. The fits appear to predict the masses of stable nuclei, and their neighbours on the mass diagram, to about 1 per cent in binding energy.

against A. To give an idea of the relative importance of the terms, for the stable nucleus (125,52) the contributions to B/A are $+ 15\cdot6$ MeV (volume), $- 3\cdot44$ MeV (surface), $- 0\cdot65$ MeV (symmetry), and $-3\cdot0$ MeV (charge). The variation with A is shown in Fig.1.3. Notice that, on the scale shown, the pairing energy term is quite negligible; for the neighbouring even-

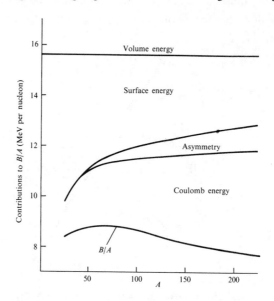

Fig.1.3. Contributions to B/A of the individual terms in the mass formula.

even nucleus to $A=125$, it contributes an additional 8 keV to B/A.

Stability against β-decay

   β^{-}-decay is represented by the reaction

$$(A,Z) \rightarrow (A,Z+1) + e^{-} + \bar{\nu}_{e},$$

where ν_e and $\bar{\nu}_e$ represent electron neutrino and elec-
tion antineutrino (see Chapter 4). This reaction is
energetically possible if

$$m(A,Z) > m(A,Z+1) + m_e \qquad (1.2)$$

since $m_\nu \sim 0$. But

$$m(A,Z) = M(A,Z) - Z\, m_e$$

leading to

$$M(A,Z) > M(A,Z+1) \ . \qquad (1.2a)$$

β^+-decay is represented by the reaction

$$(A,Z) \rightarrow (A,Z-1) + e^+ + \nu_e$$

and is energetically possible if

$$m(A,Z) > m(A,Z-1) + m_e \qquad (1.3)$$

leading to

$$M(A,Z) > M(A,Z-1) + 2m_e. \qquad (1.3a)$$

For these two types of decay it seems that the
nuclear masses produce a more attractive statement of
condition. However, there is another type of decay
- an alternative, competing, mode to β^+-emission -
namely, electron capture. Nuclei are not required to
be stable in isolation, but in their natural habitat
when surrounded by atomic electrons. Any one of these

electrons can be captured, though if energetically
possible K-capture is likely to be dominant, leaving
the atom strongly excited. However, in discussing the
limit of stability the possibility of capture of the
outermost electron, leaving the atom in its lowest
state, must be considered, since energetically it is
most favoured.

This reaction is represented by

$$(A,Z) + e^- \rightarrow (A,Z-1) + \nu_e$$

and is energetically possible if

$$m(A,Z) + m_e > m(A,Z-1), \qquad (1.4)$$

leading to

$$M(A,Z) > M(A,Z-1). \qquad (1.4a)$$

These equations indicate that electron capture
can proceed when β^+-emission is impossible and also
that the atomic mass scale is again favoured: the
condition for stability against the β-decay process is
that the *atom* (A,Z) should have least mass.

For the heavier nuclei this can be expressed,
approximately, as $(\partial M/\partial Z)_A = 0$, which is more convenient to use than the two inequalities. The corresponding condition on the nuclear mass scale
$(\partial m/\partial Z)_A = -m_e$, is probably not so convenient to remember nor indeed is the differential condition on the
binding energy, $(\partial B/\partial Z)_A = m_p + m_e - m_n$.

From these equations, two nuclei (A,Z) and
$(A,Z-1)$ can both be stable only if they have exactly

the same atomic mass. This statement is subject to approximations concerning the neutrino mass and the electron binding energies. Taking these into account might produce a band of stability a few eV wide. The chance that such a pair of nuclei should have equal atomic masses to within a few eV is remote, so it is concluded that, in the few such examples naturally occurring, one of the pair is unstable but has a very long lifetime, e.g. ^{87}Rb and ^{87}Sr - (87,37) and (87,38) - of which the former has a lifetime of 5×10^{10} years, emitting β^--particles with maximum energy $\sim$ 274 keV.

Effect of the pairing term on stability

When A is constant and odd the mass values for different Z will define discrete points lying on a parabola. For A constant but even, it is necessary to define two parabolae displaced from each other by $2\delta(A)$, which is roughly $2\cdot5$ MeV at $A = 81$, falling to ~ 1 MeV at $A = 256$. The odd-Z mass points will fall on the upper parabola and the even-Z mass points on the lower. Fig.1.4 gives examples of the parabolae for odd A and even A.

The displacement of the two parabolae for even A is sufficient to ensure, even when the lowest odd-Z mass point is at the minimum of its parabola, that both its neighbours (which will lie on the lower parabola) have lower mass. This can be checked for the two masses given above by evaluating the second derivative of the mass formula without the δ term; the mass difference between this odd nucleus, assumed to be at the minimum of the parabola, and its two even neighbours will then by $\frac{1}{2}(\partial^2 M/\partial Z^2)_A$. At $A = 81$ this is $\sim 1\cdot5$ MeV and at $A = 256$, $\sim 0\cdot6$ MeV, so that the

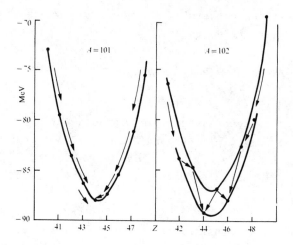

Fig.1.4.   Mass parabolae for odd A and even A. The mass excess
(in MeV) is plotted against Z for A = 101 and 102.

pairing term is amply large enough for its purpose.

These considerations lead to the conclusions that,
for odd A, only one nuclide is expected to be stable
against β-decay, but for even A two apparently stable
nuclides is likely to be a common occurrence (and is)
and even three is a remote possibility (one or two
examples are found). The term 'apparently stable' has
been used because one such stable even-even nucleus
should be able to decay to its less massive even-even
neighbour by a double β-decay process, which as yet
has not been observed.

Stability against nucleon or α-particle emission

From Fig.1.2 it is seen that for A > 20 the binding per particle for nuclei near the minima of the mass

parabolae (i.e. along the 'valley of stability' on a three-dimensional M,A,Z plot) lies between 6 MeV and 8 MeV. This large value ensures that even the most massive naturally occurring nuclei ($A \sim 240$) are stable against emission of proton or neutron.

The α-particle, on the other hand, is itself bound by $\sim$ 28 MeV, so that instability to α-emission is much more probable. The condition for instability is

$$M(A,Z) > M(A-4, Z-2) + M(4,2)$$

or

$$B(A,Z) < B(A-4, Z-2) + B(4,2) \ .$$

However, Fig.1.2 presents B/A as a function of A; if we denote this by $\phi(A)$ - along the line of stability Z is a function of A, and so $B(A,Z)$ is a function of A only - then

$$A \ \phi(A) < (A-4) \ \phi(A-4) + 4 \ \phi(4) \ .$$

But

$$\phi(A-4) \approx \phi(A) - 4 \ d\phi/dA$$

giving, after dividing by 4,

$$\phi(A) < \phi(4) - (A-4) \ d\phi/dA \ .$$

From the graph $d\phi/dA \sim 0 \cdot 75$ MeV/100 and is roughly constant well above the maximum of the curve; and

$\phi(4) \sim 7$ MeV.

Thus the condition for α-instability is that $\phi(A)$ is less than $\sim 7 \cdot 8$ MeV at $A = 100$, $\sim 8 \cdot 1$ MeV at $A = 150$, and $\sim 8 \cdot 5$ MeV at $A = 200$. α-decay is therefore possible for $A > 150$ (though shell effects have an influence; see Chapter 3). Because of the mechanism of the process (see Chapter 4) it becomes a significant mode of decay only above $A \sim 210$.

ANGULAR MOMENTUM (SPIN) AND STATISTICS

Quantum mechanically, as classically, an isolated system has a well-defined angular momentum. It is therefore a characteristic property of the nuclear ground state - and indeed any other well-defined state. As such it is loosely referred to as spin of the nucleus, even though it is a combination of the intrinsic spins and orbital motions of the individual nucleons. The reason for this is probably historical. Nuclear spins of stable and long-lived nuclei were determined from their magnetic interactions with the atomic electrons or with externally applied fields. For such interactions the spin vector is an intrinsic property of the nucleus just as it is, at present, for an elementary particle.

Since the intrinsic spin of a nucleon is $\frac{1}{2}$, whilst all orbital angular momentum vectors are characterized by integer quantum numbers, it follows that the sum of A such spins and orbital terms will lead to an angular-momentum (or spin) quantum number J which will be an integer when A is even but (integer $+ \frac{1}{2}$) when A is odd. Examination of a table of nuclear spins confirms this and also reveals a striking fact: all even-Z even-N nuclei have ground-state spin 0, without exception.

Since a pairing energy term, favouring the pairing of like nucleons, has already been introduced into the mass equation, it seems reasonable to suggest that such pairs individually couple to zero angular momentum. This point will be taken up in Chapter 3.

The Pauli principle requires that quantum-mechanical systems have wavefunctions which are antisymmetric to the exchange of like fundamental particles. If the system under consideration is atomic (for definiteness, consider a molecule made up of two like atoms) then the electronic wavefunction should have a definite symmetry on interchange of two nuclei as a whole, whilst the nuclear wavefunction should be antisymmetric on interchange of two protons (or neutrons), either within a nucleus or between two nuclei. If the like nuclei have mass number A then their interchange is equivalent to A interchanges of nucleons. Thus the electronic wavefunction will be antisymmetric for odd A and symmetric for even A. The symmetry reflects in the statistical behaviour of an aggregate of such atomic systems. Antisymmetric states conform to the Fermi-Dirac distribution, whilst symmetric states conform to the Bose-Einstein distribution (Martin 1975; Pauling and Wilson 1935); for a diatomic molecule of similar atoms, the intensities of lines in a rotational band spectrum reflect the statistics obeyed and so determine whether the number of nuclear particle components is odd or even (see Burcham 1963). This may sound trivial now, but before the discovery of the neutron it led to rejection of the hypothesis that nuclei were composed of protons and electrons. In particular, for ^{14}N (14,7) the N_2 molecule obeys

Bose-Einstein statistics, although on that theory ^{14}N
was thought to consist of 14 protons and 7 electrons -
an odd total of fermions.

The symmetry rules impinge on nuclear reactions
in a somewhat different way. Consider the system
$(\alpha+\alpha)$; its wavefunction must be symmetric under the
interchange of the two α-particles. If these particles
can come together to form a state in ^{8}Be then this
state must reflect that symmetry. But ^{8}Be is more
complex than a (2α) structure (e.g. it can also be
formed from ^{7}Li + p) and each state can be looked upon
as a conglomerate of such structures. However, of all
the states in ^{8}Be which exist only those with a certain
symmetry can decay into $\alpha+\alpha$. These states must have
$J = 0$, 2, 4, etc., as can be shown by a consideration
of the wavefunctions for $\alpha+\alpha$.

PARITY

Parity is another aspect of symmetry in quantum
mechanics, but one which has no direct correspondence
in the classical limit. If the interaction potential
of a system has the symmetry property
$V(x,y,z) = V(-x,-y,-z) = PV(x,y,z)$, where the operator
P has been introduced and defined, and the Schrödinger
wave equation has a solution $\psi(x,y,z)$, then $P\psi(x,y,z)$
is also a solution. For a non-degenerate case, the
second solution must be a numerical factor times the
first, i.e. $P\psi(x,y,z) = \alpha\psi(x,y,z)$. But it is obvious
that $P^2 = 1$, corresponding to two inversions, so that
$\alpha = \pm 1$. Thus a non-degenerate solution corresponding
to this type of potential has either even parity
$(P = + 1)$ or odd parity $(P = - 1)$.

For a single particle in a central potential there

is a one-to-one correspondence between the orbital
angular momentum and the parity, which follows from the
angular solution of the problem:

even l ≡ even parity ,   odd l ≡ odd parity.

For a multi-particle assembly like the nucleus, short
of a complete solution to the problem, the parity can
only be deduced in terms of a model in which the
parities of the individual particles can be specified,
as in the shell model (Chapter 3). From the fact that
the complete wavefunction can be expanded in terms
which are themselves products of the wavefunctions of
the individual particles, it follows that the overall
parity is a product of the individual parities if + is
assigned to even parity and - to odd parity.

As a constant of the motion, parity plays an important part in nuclear reactions. If in the reaction
$X + a \rightarrow W* \rightarrow Y + b$, where $W*$ represents a state in the
compound nucleus (Chapter 6), the spins and parities
of all five particles are known, then conservation of
angular momentum and parity severely restrict the
orbital angular momentum taken in by a and taken out
by b, thus affecting both the angular distribution
and the magnitude of the cross-section. As an example
assume in the above that X has (spin, parity) = (0^+),
a is a proton $(\frac{1}{2}^+)$, b is a neutron $(\frac{1}{2}^+)$, Y is 2^+,
the compound nucleus $W*$ is $\frac{3}{2}^-$. Then, conserving angular momentum only gives l_a = 1 or 2, l_b = 0,1,2,3, or
4, but parity conservation will reduce these possibi-
lities to l_a = 1, l_b = 1 or 3. For spinless particle
and projectile, the accessible states in the compound
nucleus are severely restricted, e.g. for $^{12}C + \alpha \rightarrow ^{16}O*$

then the excited states must be 0^+, 1^-, 2^+, 3^-, etc.

ELECTRIC AND MAGNETIC MOMENTS

Electric moments

If a charge is distributed over a finite region of space, surrounding a chosen origin, through which an electric field derivable from a potential $U(x,y,z)$ has been imposed from outside, then the interaction energy can be expressed in a Taylor series thus,

$$E = \int U(x,y,z)\rho(x,y,z)d\tau = U(0)\int \rho d\tau +$$

$$+ \left(\frac{\partial U}{\partial x}\right)_0 \int x\rho d\tau + \left(\frac{\partial U}{\partial y}\right)_0 \int y\rho d\tau + \left(\frac{\partial U}{\partial z}\right)_0 \int z\rho d\tau +$$

$$+ \tfrac{1}{2}\left(\frac{\partial^2 U}{\partial x^2}\right)_0 \int x^2\rho d\tau + \tfrac{1}{2}\left(\frac{\partial^2 U}{\partial y^2}\right)_0 \int y^2\rho d\tau + \tfrac{1}{2}\left(\frac{\partial^2 U}{\partial z^2}\right)_0 \int z^2\rho d\tau +$$

$$+ \left(\frac{\partial^2 U}{\partial x\partial y}\right)_0 \int xy\rho d\tau + \left(\frac{\partial^2 U}{\partial y\partial z}\right)_0 \int yz\rho d\tau + \left(\frac{\partial^2 U}{\partial z\partial x}\right)_0 \int zx\rho d\tau +$$

+ higher-order terms.

The first term gives the interaction energy with the field of a point charge at the origin equal to the total charge - the monopole term. The next three terms represent the scalar product of a vector dipole - the dipole moment of the charge distribution - with the vector field at the origin; whilst the next six terms represent the scalar product of two tensors, one the generalized derivative of the vector field and the other the quadrupole moment of the charge distribution. These classical definitions hold for a quantum-mechanical system if we interpret ρ as $e \sum_i \psi_i^*(r)\psi_i(r)$, where the summation i is taken over the charged particles of the system (protons in the nucleus).

The fact that a nucleus has a well-defined parity makes $\rho(\underline{r}) = \rho(-\underline{r})$. Thus every odd moment - dipole, octupole, etc. - will vanish identically. The simplest deformation from spherical results in an ellipsoid of revolution, which, if the z-axis is the symmetry axis, produces a quadrupole-moment tensor which can be specified by a single parameter Q_0 given by $eQ_0 = \int (3z^2 - r^2)\rho d\tau$ - a definition which conveniently has $Q_0 = 0$ for spherical symmetry. A positive value for Q_0 corresponds to a prolate shape for the nucleus, i.e. elongated along the symmetry axis; whilst negative Q_0 defines an oblate nucleus, i.e. flattened along the symmetry axis. As defined the dimensions of Q are those of an area for which the convenient nuclear unit is the barn (b) (1 barn = 10^{-24} cm^2 = 10^{-28} m^2). Carrying out the integral for ρ = constant within the ellipsoid gives $Q_0 = \frac{2}{5} Z(a^2 - b^2) \approx \frac{4}{5} ZR^2 \left[\frac{\Delta R}{R}\right]$, where a, b are the major and minor semi-axes, R the average nuclear radius, and $\Delta R = a-b$.

Obviously a state $J = 0$ has spherical symmetry and so has zero observable quadrupole moment; less obviously, a state $J = \frac{1}{2}$ also has spherical spatial symmetry and zero observable quadrupole moment. Table 1.1 illustrates these points and also indicates that some nuclei can be strongly deformed with $\Delta R/R \sim 30$ per cent.

Magnetic dipole moment

A similar, but more complicated, expansion of the

TABLE 1.1

Nucleus	B/MeV †	Atomic binding keV	GROUND STATE			
			J^π	μ/nm	Q/b	$\Delta R/R$ ††
n	-		$\frac{1}{2}^+$	$-1\cdot9135$	-	
$^{1}_{1}$H	-	$0\cdot014$	$\frac{1}{2}^+$	$+2\cdot79275$	-	
$^{2}_{1}$H	$2\cdot225$	$0\cdot014$	1^+	$+0\cdot85735$	$+0\cdot00282$	?
$^{4}_{2}$He	$28\cdot29$	$0\cdot079$	0^+	0	0	0
$^{7}_{3}$Li	$39\cdot24$	$0\cdot190$	$\frac{3+}{2}$	$+3\cdot2564$	$-0\cdot04$	?
$^{16}_{8}$O	$127\cdot6$	$2\cdot0$	0^+	0	0	0
$^{35}_{17}$Cl	$298\cdot2$	12	$\frac{3+}{2}$	$+0\cdot8218$	$-0\cdot08$	$0\cdot1$
$^{57}_{26}$Fe	$499\cdot9$	34	$\frac{1}{2}^-$	$+0\cdot0905$	0	0
$^{121}_{51}$Sb	1026	180	$\frac{5+}{2}$	$+3\cdot342$	$-0\cdot53$	$-0\cdot12$
$^{176}_{71}$Lu	1418	370	$\frac{7-}{2}$	$+3\cdot180$	$+8\cdot0$	$+0\cdot50$
$^{181}_{73}$Ta	1452	390	$\frac{7+}{2}$	$+2\cdot35$	$+3\cdot9$	$+0\cdot35$
$^{209}_{83}$Bi	1640	550	$\frac{9-}{2}$	$+4\cdot080$	$-0\cdot34$	$-0\cdot02$
$^{238}_{92}$U	1783	690	$\frac{7-}{2}$	$\pm0\cdot35$	$\pm4\cdot1$	$\pm0\cdot25$

Table 1.1 continued

Footnotes: † B is given only to four figures; it is
     however known to the nearest keV for the light
     elements and somewhat less precisely for the
     heavy.
          †† The $\Delta R/R$ have been computed using
     $R = 1 \cdot 1 A^{1/3}$ fm. A relationship
     $Q/Q_0 = J(2J-1)/(J+1)(2J+3)$ is used to connect
     the observed quadrupole moment Q with the in-
     trinsic quadrupole moment Q_0.

charge distribution and its interaction with an external magnetic field, taking into account a possible
circulation of the charge within the distribution will
give a leading term dipole in nature. If in transition
to the particle system the intrinsic magnetic moments
of the individual particles are taken into account,
this dipole term will have two parts: (1) from the
circulating currents of the proton distribution which
will be proportional to some suitable summation of the
orbital angular momenta l_i and (2) from a suitable
summation of the intrinsic magnetic moments of both
protons and neutrons which will be vectors parallel
to the spins s_k. The phrase 'some suitable summation'
is used because the resultant magnetic moment of a number of particles depends on how the particles are
coupled together; recourse to a model is necessary,
as will be apparent in Chapter 3. At this stage it
should be noted that the magnetic dipole moment is a
vector, and it is a fundamental result in quantum
mechanics that all vector constants of a system are
scale factors of one vector, which, for an isolated

system, will be the total angular momentum J. Thus
vectorially $\underline{\mu} = g\underline{J}$, where g is the suitable scale
factor. It is convenient to represent the dipole
moment by a scalar, which in the classical limit would
represent the magnitude $|\mu|$. Two possibilities exist:
$g\{J(J+1)\}^{\frac{1}{2}}$ and $g(m)_{max} = gJ$. The latter has been
adopted.

For the case of a spinless particle of mass m and
charge e, J = L and g is the gyromagnetic ratio $e/2m$ -
a relationship which can readily be verified for a circular orbit, using Ampere's theorem connecting a circulating current with a magnetic shell. This relationship leads to a natural unit $e\hbar/2m$ for the dipole moment. In atoms the mass is that of the electron, defining the Bohr magneton (μ_B); in nuclei it is the
mass of the proton defining the nuclear magneton (μ_N),
which has the value $5 \cdot 051 \times 10^{-27}$ J T^{-1}. The intrinsic
magnetic dipole moments of the proton and neutron are
respectively +2·79275 μ_N and -1·9135 μ_N. A few nuclear
magnetic moments are given in Table 1.1.

PROBLEMS[†]

1.1. Assuming that the nuclear size parameters are as
    on p. 12, calculate the quantum number n for which
    the circular Bohr orbit of K$^-$ around $^{208}_{82}$Pb lies
    in nuclear matter at a density of 10 per cent of
    the central nuclear density. The mass of the K$^-$
    is 494 MeV/c^2.

1.2. By assuming that the neutrons and protons in the

[†] * indicates that the problem is taken from, or
based on, an Oxford Finals question.

nucleus behave as independent Fermi gases, and
by neglecting the Coulomb interaction, derive
the form of the symmetry term, $\propto (N-Z)^2$, which
appears in the mass formula.

1.3. Verify the statement that the pairing term in the
mass equation is sufficiently large to ensure
no odd-odd nucleus (above $A=40$) is less massive
than both its even-even neighbours of the same
mass number.

1.4*.Calculate values (in MeV/c^2) for γ and ε in the
mass equation from the following facts: $^{35}_{18}\text{A}$ emits
positrons with a maximum energy of 4·95 MeV;
$^{135}_{56}\text{Ba}$ is the stable isobar of mass number 135.

1.5*.The maximum energy of the positrons from the
decay of $^{13}_{7}\text{N}$ is 1·24 MeV and there is no subsequent γ-radiation. Deduce a value for the
radius of nuclei of mass 13. (The n-p mass difference is 1·29 MeV/c^2.)

1.6. Show that the angular wavefunction $\sin\theta\ \exp(i\phi)$,
corresponding to $l=1$, $m_l=1$, has odd parity,
whilst $\sin^2\theta\ \exp(i2\phi)$ and $\sin\theta\ \cos\theta\ \exp(i\phi)$,
corresponding to $l=2$, $m_l=2$ and 1, have even
parity.

1.7. Deduce the gyromagnetic ratio for the case of a
charged particle moving in a circular orbit.
Hence derive the natural units for the magnetic
moments of electron and proton orbits.

2. Nuclear forces

INTRODUCTION

From the last chapter a picture of the nucleus has
emerged which is quite different from that of an atom.
In the latter, the small massive central nucleus interacts more strongly ($\sim Z$ times) with any electron than
that electron does with any other electron. For the
nucleus there is no well-defined centre; the interaction of one nucleon with its neighbour at any instant
of time is probably stronger than its interaction with
the nucleus as a whole - this latter being a time-
averaged effect of all the individual interactions.
Thus the problem is a many-body one, without the obvious approximations of the atomic case; in the next
chapter models will be illustrated which simplify the
many-body aspects. The shell model has been taken
over from the atom, but initially with reluctance,
since its basic concepts seem to clash with ideas of
the nucleus.

In this chapter, the basic nucleon-nucleon interaction is examined in order to answer the questions:

(1) Assuming that the strong interaction between nucleons can be represented by a potential function,
what is its basic shape and does it depend on the
type of nucleon?

(2) Arguing backwards from a nucleus towards a basic
nucleon-nucleon interaction, does this interaction agree with that deduced from basic nucleon-
nucleon scattering experiments?

(3) Knowing that the answer to (2) is indefinite, is there a need to introduce new forces in the nucleus which are not present in the basic two-body interaction?

The answer to the third question can be given now. Theorists are not happy with the idea of many-body forces, although there appears to be no very good reason why they should be small compared with the two-body interaction. The subject has not yet reached such a stage of refinement as to need a small 'topping up' with three-body forces. If they are to be introduced at all, it may as well be done in bulk - in which case what of four-body forces, etc? The hope appears to be that averaging many-body forces in a large nucleus produces much the same effect as averaging two-body forces. Note that, in answering this question, addition of a third nucleon will always modify the interaction between two nucleons because of the Pauli principle - but this effect is always allowed for in any attempt to solve the problem of larger nuclei and is not to be looked upon as a new three-body interaction in the sense referred to here.

INFERENCES FROM THE DATA PRESENTED

In developing the nuclear properties of the previous chapter, a number of facts have emerged which have a direct bearing on the forces which hold the nucleus together. Taken roughly in the order in which they were presented they are:

(1) nuclear sizes are of the order of a few fm;

(2) nuclear density, at least near the centre, is the same for all nuclei;

(3)   protons and neutrons have similar density dis-
      tributions inside the nucleus;

(4)   scattering of nuclear projectiles gives the
      Rutherford cross-section up to a certain angle
      with a rapid fall-off above this angle;

(5)   the basis for a successful analysis of electro-
      static energy differences between mirror pairs
      is that nuclear forces are 'charge symmetric',
      i.e. pp ≡ nn;

(6)   the nuclear density falls off over a range of
      ∿ 1 fm at the periphery;

(7)   the leading terms in the binding energy,
      $\alpha A - \beta A^{2/3}$, were arrived at by analogy with a
      liquid drop;

(8)   there is a 'symmetry term' in the mass equation;

(9)   the existence of a pairing term, and the coupling
      to zero angular momentum of every ground state
      of even A indicate a pairing force;

(10) the α-particle is almost as tightly bound as a
      massive nucleus;

(11) some heavy nuclei have large quadrupole moments;

(12) even the deuteron has a quadrupole moment;

(13) the proton and neutron have 'anomalous' magnetic
      moments, i.e. different from those for structure-
      less Dirac particles;

(14) the magnetic moment of the deuteron is close to,
      but not equal to, the sum of the proton and
      neutron magnetic moments.

RANGE AND SATURATION

The terms 'long-range' and 'short-range' have
both a general and a specific meaning. In the general
sense long and short are relative to the size of the
system; van der Waals forces of cohesion of a liquid
drop are short-range in the sense that the binding of
a given molecule is dominated by the contributions of
a few fairly close neighbours rather than by the overwhelmingly greater number of distant molecules - these
forces $\propto 1/r^7$ (or so) in energy-dependence. Likewise,
in nuclei, forces are short-range if the contribution
of most nucleons to the binding of a given one is less
than the contribution from its near neighbours. In
the specific sense, forces are said to have a range
if the potential function contains an exponential, e.g.
Yukawa forces having a potential function
$U \propto (1/r)\exp(-kr)$ are said to have a range $1/k$. Thus
it will be seen that nuclear forces probably are short-
range in both the general and specific senses, though
arguments on saturation are concerned really with the
general sense.

The nuclear force field extends beyond the nuclear surface by a distance of the order of magnitude
of the range of nuclear forces, so (1) and (4) (of the
previous section) taken in conjunction can be used to
define this range. The estimate obtained is confirmed
by (6) since we would, in a simple way, expect the
fall-off of nuclear density to occur over the same
range.

(2) and (7) in conjunction with the liquid-drop
model from which the mass terms were derived do much
to define nuclear forces. The molecular force field
in a liquid drop has two main components: the attrac-

tive short-range force, which can be oversimplified by
describing it as an induced-dipole - induced-dipole
interaction for which the potential energy $\propto r^{-6}$; and
the repulsive, even shorter-range force which occurs
when the electron clouds overlap. By analogy, a similar state of affairs may be expected to occur in nuclei;
the short-range attraction would require a range of
1 - 2 fm, whilst the shorter-range repulsion might have
a range of less than 0·5 fm.

Oddly enough, in the historical development of
the subject, the concept of a repulsion at short distance was not readily acceptable. Nucleon-nucleon
scattering had not yet indicated the need for such a
force, which appeared to require a complex structure
for what was hoped to be a simple fundamental particle.
Instead, recourse was made to the exchange phenomenon
(see Appendix C) to provide forces with suitable properties. Such forces were described as: Wigner (W)
or ordinary (no exchange of quantum numbers); Majorana
(M) or space-exchange (corresponding to exchange of
charge and spin); Heisenberg (H) or charge-exchange
(but not spin); and Bartlett (B) or spin-exchange (but
not charge). All these forces can arise from pion-exchange. From analogy with electron-exchange binding
of molecules, a nucleon should be surrounded by a pion
cloud. That this is so can be inferred from the magnetic moments of proton and neutron, which for simple
Dirac particles should be 1 μ_N and 0 μ_N; but the pion
has a much smaller mass than proton and neutron so the
cloud can, and does, make a large contribution to their
moments.

Exchange forces will depend upon whether the property being exchanged is symmetric or anti-symmetric

in the wavefunction representing the relative motion
of the two particles. As an example the spin-exchange
operator (see Appendix C), $P^{\sigma} = \frac{1}{2}(1+4\underline{s}_1 \cdot \underline{s}_2)$, has the
value +1 for $S = 1$ and -1 for $S = 0$, where
$\underline{S} = \underline{s}_1 + \underline{s}_2$, so the spin-exchange force can be made
attractive for $S = 1$ and repulsive for $S = 0$, or vice
versa.

How do exchange forces lead to saturation? In
a heavy nucleus A, the number of nucleon-nucleon interactions is $\frac{1}{2}A(A-1)$. If the forces are ordinary then
all these are increasing the binding of the nucleus,
and a complete calculation leads to all nuclei having
roughly the same size with a binding energy proportional to A^2 rather than A. However, the operation
of the Pauli principle ensures that all nucleon pairs
cannot have the same spin symmetry or charge symmetry or spatial symmetry, so that exchange forces
will as often be repulsive as attractive. Their overall effect need not be zero, since the strength of
the force will depend upon a spatial overlap integral
which of course is largest when the spatial wavefunctions of the two particles are the same. The exclusion principle, in its restricted statement, limits
the number of particles with the same spatial wavefunction to 4 - two protons (spin up, spin down) and
two neutrons. Thus Majorana (spatial) exchange forces
might be expected to produce strong interactions within groups of four particles and weak effects on averaging over pairs not in the same group. The emergence of four particles is an attractive result because (10) emphasizes the fact that the α-particle is
about as strongly bound as any heavy nucleus.

Much effort has gone into producing a recipe for

nuclear forces using these four types - but each recipe, whilst successful for explaining the particular
fact on which the 'cook' was concentrating, has had
its shortcomings, and the fact of a repulsive core
(see p. 54) has been finally accepted with relief.
This does not mean that exchange forces are unnecessary;
it is difficult to see how ordinary forces with repulsion could lead to a saturated α-particle. Indeed,
from the range of the repulsive force as required from
nucleon-nucleon scattering, it is possible to attach
an effective size to each nucleon ($r_{\text{eff}} \sim \frac{1}{2} \times$ range,
since two nucleons cannot approach closer than this
range). The effective volume of the nucleons in a
heavy nucleus is then found to be ~ 1 per cent of the
nuclear volume. At its face value this indicates that
saturation is largely accounted for by exchange forces,
though no doubt the repulsive core also plays an important role.

ISOSPIN

In (5) (see p. 38) the point is made that forces
are 'charge symmetric', by which it means that the
(p,p) and (n,n) interactions are the same, apart from
the electromagnetic contribution, but the (p,n) could
be different. In the analysis of 'mirror paris' the
numbers of (p,p) and (n,n) interactions are inter-
changed but the number of the nuclear-force binding
does not specify the last interaction.

A further step to 'charge independence' is made
by assuming that all three interactions are the same;
this can be checked by analysing nucleon-nucleon
scattering or by a more detailed examination of nu-

clear properties. Charge independence is expressed
mathematically using the isospin formalism, which was
developed as a convenience before it acquired physical
significance. Rather than referring to two distinct
types of particle within a nucleus, theorists preferred to employ a nucleon to which an additional two-
valued quantum number was assigned and given the value
$+\frac{1}{2}$ for the neutron and $-\frac{1}{2}$ for the proton. The reason for this choice is that all two-valued quantum
numbers have a similar mathematical behaviour, so the
values reflect the similarity with $S_z = \pm\frac{1}{2}$ for the
spin quantum number.

The name 'isospin' is a shortening of 'isobaric
spin', indicating that the two components have the
same mass and emphasizing the similarity with spin S.
By analogy, there exists a vector $\underline{t}$ and a component
t_z of isospin, but the space in which this vector
exists is not the same as the space in which ordinary
spin exists. Thus, in this space, $t_z = +\frac{1}{2}$ represents
a neutron and $t_z = -\frac{1}{2}$ a proton.[†] Pursuing the analogy,
a vector sum $\underline{T} = \underline{t}_1 + \underline{t}_2$, corresponding to $\underline{S} = \underline{s}_1 + \underline{s}_2$,
can be made for two (and more) nucleons. A nucleon
pair can have $T = 1$ giving a function symmetric on
exchange of nucleons, or $T = 0$, antisymmetric. If
the Pauli principle is extended to make the overall
wavefunction antisymmetric with the isospin component
included then it can be seen that the same states in

[†]It is unfortunate that particle physicists have
later chosen precisely the opposite convention - in
principle it does not matter, since the direction up
or down in an unspecified space is hardly relevant,
but it is confusing.

the two nucleon configuration are obtained whether or
not this formalism is used; since (p,p) and (n,n) are
symmetric in isospin their wavefunctions must be antisymmetric in (space) (spin), but the (n,p) combination
can be symmetric or antisymmetric in isospin ($T = 1$,
$T_z = 0$ represents (p,n) and is symmetric; $T = 0$,
$T_z = 0$ also represents (p,n) and is antisymmetric) so
the (space) (spin) wavefunctions can also be either.

Thus far, the formalism is merely mathematical,
but charge independence can readily be fed into the
scheme by asserting that nuclear forces depend only on
T and not on T_z. This achieves its aim because the
selection of T determines the type of (space) (spin)
wavefunction and so ensures that we compare like with
like when testing the similarity of (p,p),(p,n), and
(n,n) interactions. The equality of strong interactions in mirror pairs of nuclei follows from this
statement, at least for $|N-Z| = 1$, because these mirror
pairs have $T_z = \pm \frac{1}{2}$ and, moreover, have the same value
for T, also $\frac{1}{2}$. Extending to the case $|N-Z| = 2$; ^{14}C
and ^{14}O, corresponding to interchange of neutrons and
protons, have the same value for $T(= 1)$, but $T_z = \pm 1$.
But charge independence further requires that there
should exist a state in ^{14}N having $T = 1$, $T_z = 0$ with
similar strong-interaction energy as the ground states
of ^{14}C and ^{14}O. The same should apply to excited
states in ^{14}C and ^{14}O. The state of affairs is shown
in Fig.2.1, where, to emphasize the point, the nuclei
have been moved relative to each other through energies to allow for the electrostatic term and the p-n
mass difference. An interesting feature is that the
ground state of ^{14}N has $T = 0$; there is a general tendency for states of lower T to lie lower which re-

sults in the symmetry term in the mass equation (point (8) of p. 38). The method used to derive it implied charge independence since it treated neutrons and protons in a similar way, which again is implied in point (3).

Thus another quantum number has been introduced into the description of a nuclear state but, as the shift of energy in Fig.2.1 has perhaps indicated, it

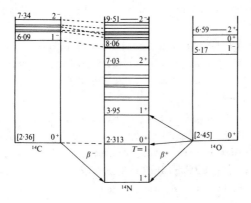

Fig.2.1. Comparison of energy levels in ^{14}C, ^{14}N, and ^{14}O. Allowance has been made for the Coulomb energy and p-n mass difference. Note that ^{14}C and ^{14}O have $T = 1$ states only, whereas ^{14}N has $T = 1$ states and a greater number of $T = 0$ states. The dotted lines guide the eye to $T = 1$ states in ^{14}N. (Based on AJZENBERG - SELOVE (1970). *Nucl.phys.* A152, 1.)

is only an approximate one, since the electromagnetic term, whilst splitting the isospin sub-states by a large energy (several MeV), is also of a form which mixes states with different T to some (small) extent.

This quantum number plays a significant role in nuclear reactions; e.g. the reaction $^{16}O + d \rightarrow {}^{14}N + \alpha$ leaves ^{14}N in excited states which are characterized by $T = 0$, since ^{16}O (ground state), d, and α all have $T = 0$. Since the state at 2.31 MeV belongs to the $T = 1$ triplet which includes the ground states of ^{14}C and ^{14}O, it can be populated in this reaction only by way of the small impurity in the wavefunction of the $T = 0$ component. This selectivity in (d,α) and similar reactions had been noticed experimentally and led to the conclusion that T was indeed quite a good quantum number.

Finally, to return to the mass equation, the pairing term (point (9)) appears to refute charge independence in favour of charge symmetry, since the term increases the binding for (p,p) and (n,n) pairs but decreases it for (p,n). However, we are failing to compare like with like; the proton pair, or neutron pair, have similar quantum numbers - this will be more obvious after reading about the shell model in the next chapter - whilst in a large nucleus the (p,n) pair are in dissimilar orbits. However, in the very light nuclei when the (p,p),(n,n), and (p,n) interactions all refer to nucleons in the same orbits (shells), it is interesting to note that the odd-odd nucleus is stable on four occasions. Thus the pairing term can be shown to be consistent with charge independence.

THE DEUTERON AND THE VIRTUAL DEUTERON

Since the deuteron presents a simple two-body problem, its study was hoped to reveal the properties of the nucleon-nucleon interaction. In hindsight it has proved to be somewhat of a disappointment; the solution

for any simple short-range potential function which
might approximate to the nucleon-nucleon interaction,
e.g. the square well having potential - V_0 from $r = 0$
to b and zero for $r > b$ (see Fig.2.2) indicates that,

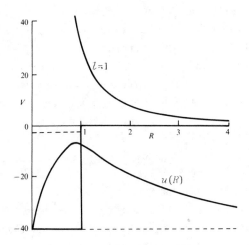

Fig.2.2. Typical square well for the deuteron, depth 40 MeV, size
1·6 fm (the scale factor for the abscissa). Also shown
is the centrifugal barrier for $l = 1$, and $u(r) = rF(r)$,
where $F(r)$ is the radial wavefunction of the ground
state of the deuteron.

in the loosely bound deuteron, the nucleons are for
most of the time outside the effective range of the
force - farther apart than b in the square well quoted
above. To pursue this example it turns out that the
wavefunction can be very similar for a large range of
variation of the square-well parameters, provided that
such variation keeps $V_0 b^2$ approximately constant. The
value of the constant is $\sim 10^{-24}$ MeV cm^2 i.e. 1 MeV b

corresponding to something like $V_0 \sim 50$ MeV, $b \sim 1\cdot4$ fm
This condition follows from that fact that $Kb \sim \frac{1}{2}\pi$,
where K is the wavenumber inside the potential, for
successful matching of the internal and external wavefunctions to produce a stationary state. For a more
realistic potential without discontinuities, a similar condition can be deduced in the form of a phase
integral in the region of positive kinetic energy. It
need hardly be added that the analysis of the deuteron
sheds no light on the existence of a repulsive core.
Also, because of the fact that the deuteron size stems
mainly from that part of the wavefunction corresponding
to negative kinetic energy and having a (slow) exponential fall-off, feeding into the problem the (mean)
size from electron scattering does not serve to pin
down the force range with any precision.

An interesting result arises, however. The lowest
state is undoubtedly an S-state. That this state is
bound only by $2\cdot2$ MeV, in a well of depth ~ 50 MeV,
makes no other bound state possible; the next S-state
with a node in the wavefunction requires a wavenumber
of approximately 3 times that of the present state,
which is clearly impossible, and the lowest $l = 1$
state is made unbound by the 'centrifugal potential'.
Classically, a circular rotor with angular momentum
$m\omega r^2$ has kinetic energy $= \frac{1}{2}m\omega^2 r^2$, which becomes
$l(l+1)\hbar^2/2mr^2$ on quantizing the angular momentum to
$\{l(l+1)\}^{\frac{1}{2}}\hbar$. A term of this form automatically arises
in the radial Schrodinger equation - it is shown in
Fig.2.2 for $l = 1$. Numerically, to bind a state having
$l = 1$ would require a well-depth of about 200 MeV for
the same range. It is highly unlikely that a mixture
of exchange forces could be produced capable of such

large variation with l of the potential.

However, it is known from low-energy neutron scattering by protons that there exists a virtual state in the deuteron unbound by $\sim$ 30 KeV. The detailed analysis of scattering is beyond the scope of this discussion. Suffice it to say that from measurements of the differential cross-section it is possible to deduce the presence of resonances (or virtual states) and to determine the orbital angular momentum l taken in by the neutron in forming this virtual state. The present experiments were performed at thermal energies and could only detect such a resonance for l = 0 (see later in chapter). Thus the virtual state is also an S-state.

For the simple potential $V(r)$ there should be spin-degeneracy giving two S-state wavefunctions, namely, J = 1, ${}^3S_1$, and J = 0, ${}^1S_0$. That they are not degenerate is proof of the spin-dependence of nuclear forces and therefore of exchange - but the analysis does not lead to a recipe for the mixture. The ground state is obviously spatially symmetric and spin symmetric; in isospin it must be the antisymmetric T = 0 state. The virtual state is spatially symmetric and spin antisymmetric; in isospin it must be the symmetric T = 1, T_z = 0 state. From the operator forms of the different exchanges (see Appendix C) it is seen that the Majorana term should be the same for both states, but that both Heisenberg and Bartlett terms should change. In the absence of any other states of the deuteron system, no evidence is available on the Majorana term; the effect of the other two terms would appear to be small, but the overall picture is far more complex than presented here so no quantitative conclusions are drawn.

So far, a simple theory has assigned the characteristics $^3S_1$, $J = 1$, parity even - denoted by 1^+ -
to the ground state of the deuteron. An S wavefunction is necessarily symmetric in space, so the deuteron should have no quadrupole moment. But, as stated
in point (12) (see p. 38), the deuteron does have a
(small) quadrupole moment. There can be no doubt that
1^+ is correct since it is soundly based experimentally.
The only other possible wavefunction having 1^+ is the
$^3D_1$-state, i.e. the coupling of spin 1 to $L = 2$ ($L = 1$
has odd parity) to give $J = 1$. This wavefunction
cannot by itself form the ground state since it has already been shown that no state possessing orbital
angular momentum can be bound by any reasonable
potential, but in any case it would give too large a
quadrupole moment. The remedy is to mix a small amount
($\sim$ 4 per cent probability) of this wavefunction with
the basic $^3S_1$ wavefunction. But this mixture no longer
has a well-defined value for L; L is certainly a
constant of motion for any central force, and so a non-
central term must be included in the potential. Such
an interaction is already present. The two nucleons
have magnetic dipole moments giving a dipole-dipole
interaction of the form

$V_{dd} \propto r^{-3}\{3r^{-2}(\underline{\mu}_1 \cdot \underline{r})(\underline{\mu}_2 \cdot \underline{r}) - (\underline{\mu}_1 \cdot \underline{\mu}_2)\}$, i.e.

$\propto r^{-3}\{3r^{-2}(\underline{s}_1 \cdot \underline{r})(\underline{s}_2 \cdot \underline{r}) - (\underline{s}_1 \cdot \underline{s}_2)\} = r^{-3} S_{12}$. This
magnetic interaction is, however, quite negligible -
something of the order of a few hundred eV. It is
therefore necessary to introduce a strong interaction
of the form $V(r) S_{12}$, where $V(r)$, some radial function
with a range, replaces the electromagnetic function
r^{-3}. This term is obviously different for the parallel

spin orientations of Fig.2.3. Since the deuteron is
prolate, the left-hand configuration must lead to an
attractive force, giving a negative sign to the potential term above. Note that the large quadrupole moments of heavy nuclei (point (11)) result from correlated motions of a number of nucleons and can arise
with central forces between nucleons.

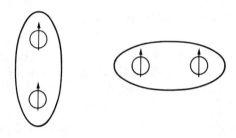

Fig.2.3. Schematic representation of the spin-spin interaction
         in the deuteron. The spins are shown directed along the
         symmetry axis.

This admixture of $^3D_1$ in the wavefunction is also
required in accounting for the magnetic dipole moment
of the deuteron. As stated in point (14), the magnetic moment is almost equal to the sum $\mu_p + \mu_n$; for
a pure $^3S_1$-state it should be exactly equal. Classically this statement is obvious; quantum mechanically
it is much less obvious. A calculation along the lines
of the Landé approach to hyperfine structure in atoms
(see e.g. Kuhn 1961) is needed to produce this result and also to determine the magnetic moment of the
$^3D_1$-state. Adding the few per cent admixture of $^3D_1$
moves the magnetic moment in the right direction.
At this stage a note of caution should be injected.

If the magnetic moments of nucleons derive in large
part from the meson clouds, what happens when nucleons
are in proximity? For heavy nuclei this could lead
to difficulty, but for the deuteron the nucleons are
never really close enough for it to matter.

From the fact that the admixture of $^3D_1$ is small,
it would be hasty to conclude that the non-central
part of the potential is also small compared to the
central terms. The 'centrifugal potential' for the
D-state for such a small system serves to reduce its
effectiveness to such an extent that it becomes important to keep the range of the tensor force at least
as large as that of the rest of the potential, in
order to keep its magnitude down to an acceptable
value. Both the exchange central force and the tensor
force can be derived from pion-exchange. The fact
that the pion has spin 0 but negative parity means
that it must exchange with $l = 1$ in order to preserve
nucleon parities. This transfer of orbital angular
momentum can cause the spin-flip necessary in some
exchange terms and can also produce the tensor term.
The exchange of a pion pair (0^+) results in predominantly S-wave transfer and an ordinary force (the
combination of two pions gives $T = 2$ or $T = 0$ symmetric and $T = 1$ antisymmetric; these must combine
with a symmetric and an antisymmetric space function
respectively. Thus $T = 0$, $S = 0$, $L = 0$ is the only
possibility for exchange). Since it is also repulsive it is believed to be (partly) responsible for the
repulsive core.

NUCLEON-NUCLEON SCATTERING

The (n,p) scattering system at thermal neutron

energies has been cited (p.49) as a source of information on the $^1S_0$ virtual state of the deuteron.
Scattering at higher energies with this system and
the (p,p) system gives much more information provided that the energy is high enough. Wavemechani-
cally, the projectile is represented by a plane-wave
passing the target, placed at the centre of the coordinate system. Such a plane-wave can be analysed
into an infinite series of spherical waves converg-
ing on and diverging from the coordinate centre and
characterized by the value of the orbital angular
momentum l. This series is the quantum-mechanical
equivalent of the classical sub-division of the wave-
front into zones characterized by the impact parameter of the collision; if this latter parameter is
p then $pmv = \hbar \{l(l+1)\}^{\frac{1}{2}}$ is the correspondence, where
v is the velocity of the projectile of reduced mass
m. It is interesting to put numbers into this equation, using the nucleon-nucleon system to define m:
for $l = 1$, 20 MeV incident neutrons in the laboratory
system will correspond to a classical impact parameter of 3 fm. The $l = 1$ partial wave of the plane
wave corresponding to 20 MeV neutrons will therefore
be unaffected by the collision if the interaction is
weak at a distance of 3 fm (the corresponding distance
at E_n = 1 MeV is 13 fm). Thus there is a need for
higher and higher projectile energies in order to
determine the interaction for higher l-values. Although thermal neutrons were sufficient for the S-wave
interaction, they could give no information on the
interactions at higher l-values for which laboratory
energies $\sim$ 30 MeV are necessary. At $E_n \sim$ 300 MeV,
the impact parameter is down to 0·7 fm for $l = 1$, and

so information can be obtained on the potential at
close approach.

Note that this same equation with $\{l(l+1)\}^{\frac{1}{2}} = 1$
is also an expression of the uncertainty principle;
if the wavelength in the plane wave is λ, then this
distance is also the measure of the uncertainty with
which a particle can be located and is therefore a
limit to which the system can be investigated. Thus
for the S-wave interaction an energy of $\sim$ 300 MeV is
also required to probe the repulsive core. It is from
experiments performed at these energies that the need
arose for a repulsive core in the nucleon-nucleon
interaction.

Another result can be discerned from a qualitative
examination of neutron-proton scattering (see Fig.
2.4). The scattering in the centre-of-mass system

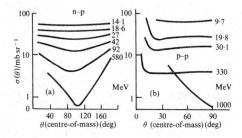

Fig.2.4.   (a) n-p scattering at high energies.   (b) p-p scatter-
            ing at high energies. Note, because of similarity
            of striking and struck particle, the angular distri-
            bution must be symmetrical about 90° (centre-of-mass).
            (Based on HESS (1958). *Rev.mod.Phys.* **30**, 368.)

rises sharply toward 0° and 180° at high bombarding
energies. Peaking at 0° is to be expected; scattering
through large angles requires a head-on collision (or
nearly so), whilst small-angle scattering can occur
when the neutron passes the proton at a distance and
so experiences the extreme tail of the potential.
Many more neutrons will 'miss' the proton than 'hit'
it. This näive picture conforms with analysis provided the potential is simply $V(r)$, i.e. an ordinary,
central potential. The rise at 180° is proof of
charge-exchange, corresponding to a small-angle de-
flection coupled with exchange of charge; the detected
neutron, which was originally the proton, moves in
the centre of mass in the opposite direction. The
type of exchange has been described as charge (or
Heisenberg); it could equally have been charge-spin
(or Majorana), since the measurements shown refer to
unpolarized target and projectile and therefore are
averaged over spins. As has been stated previously
a large Majorana term is already required.

Another point arising from nucleon-nucleon scattering is that the charge-independence hypothesis
appears to be wellfounded. This is not a result which
can be arrived at qualitatively; indeed the qualitative conclusion (see Fig.2.4) might well be the
opposite. The (p,p) system is, of course, affected
by Coulomb forces, and there is an added complication
that the two particles are identical, giving rise to
interference terms on correctly antisymmetrizing the
wavefunction. Even when these differences are taken
care of, the comparison of (p,p) and (n,p) is still
not like with like: the (p,p) system must have
$T = 1$ whilst the (n,p) has equal admixtures of $T = 1$

and $T = 0$. The analysis is lengthy and complicated, but the final result appears to be that the $T = 1$ component of (n,p) behaves similarly to the (p,p) when allowance is made for Coulomb and symmetry effects.

One final piece of information has arisen from nucleon-nucleon scattering with polarized beam and target. A spin-orbit coupling term ($\underline{L}.\underline{S}$) appears to be necessary. Again it has a magnetic analogue, but again it is far too weak. In introducing a strong interaction of this type, the radial function has been given a short range since the term seems to become more effective at higher energies.

SUMMARY

Nuclear forces have been shown to have:

(1) a short range, of order 1·5 fm;

(2) a considerable exchange component, largely of the Majorana type;

(3) a repulsive core, of range $\sim$ 0·5 fm;

(4) a tensor component;

(5) a velocity-dependent spin-orbit interaction.

The greater part of these requirements can be met by a potential derived from exchange of a pion, giving a range of 1·4 fm and accounting for the exchange and tensor components. The repulsive core can be accounted for by simultaneous exchange of two (or more) pions or by exchange of heavier mesons. Among these heavier mesons must be included a vector meson ($J^{\pi} = 1^{-}$) in order to account for the spin-orbit term.

A rough qualitative picture of the nuclear potential is given in Fig.2.5. Notice that in detail

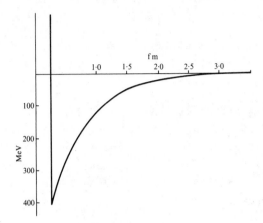

Fig.2.5. An indication of the nucleon-nucleon potential, with
hard-core and pion-exchange forces. Note that, in
detail, the potential depends upon the combination
of T,L,S,J.

it should be different for different combinations of
the quantum numbers L, S, and T.

PROBLEMS

2.1. If the n-p potential is of the form
$V_1(r) + V_2(r)\underline{s}_1 \cdot \underline{s}_2$, show that J^2, parity, L^2,
and S^2 are constants of the motion.

2.2. Determine the magnetic moment of the deuteron
in the $^3D_1$-state and show that a few per cent
admixture of this state into the $^3S_1$-state gives
the measured magnetic moment.

2.3. Show that the formal introduction of isospin together with a generalization of the Pauli prin-

ciple does not affect the number of states
available to two nucleons.

2.4. The Serber potential is of the form $V(r)(1+P^\sigma)$
(see Appendix C). Show that it gives no force
in odd spatial states.

2.5*. The ground states of $^{10}_4\text{Be}$ and $^{10}_6\text{C}$ lie respectively 0·56 MeV and 3·58 MeV above the 3^+ ground
state of $^{10}_5\text{B}$. Estimate the energy and spin of
one excited state of $^{10}_5\text{B}$. Comment upon the possibility of exciting this state by deuteron inelastic scattering.

2.6*. Discuss what facts you can deduce about the
interaction between two nucleons from the following observations:

(a) in the elastic scattering of neutrons by protons
the angular distribution of the scattered neutrons is nearly isotropic in the centre-of-mass
system for incident energies below 10 MeV;

(b) when the incident energy is raised above 100 MeV,
sharp peaks develop at 0° and 180°;

(c) the binding energy per nucleon in medium-weight
nuclei is approximately constant and independent
of mass number;

(d) the cross-sections for scattering of slow neutrons by ortho- and para-hydrogen differ by a
large factor.

3. Nuclear models

In the previous chapters, simple nuclear models have been introduced, namely, the liquid-drop and Fermi-gas models, which are in effect classical and quantum aspects of the same system. This chapter will discuss three models: the shell model, the collective model, and the optical model. The first is an extension of the Fermi-gas model, whilst the last is a dynamical model concerned with the nucleus at highish excitation, where the restriction of the Pauli principle is decreasing in importance and the system is undergoing transition to something like a liquid drop. In the collective model the restriction to spherical symmetry is removed and the nucleons correlate their motion in the well in such a way as to create the non-spherical well in which they are moving.

THE SHELL MODEL

The atomic shell model has had a great deal of success in accounting for the construction of atoms and the chemical properties of the elements. A brief description is therefore appropriate.

Solution of the Schrödinger equation for the simple system of one electron bound by a central nucleus of charge Z units results in energy levels characterized chiefly by the principal quantum number n. Neglecting the fine structure, each level is degenerate having orbital angular momenta $l = 0, 1, \ldots (n-1)$, which, coupled with spin degeneracy 2, gives a total degeneracy $2n^2$. Also the scale factor of the radial wavefunction

is largely determined by n. In filling up such an
atom with its Z electrons, account must be taken of
the Pauli principle. Thus the first two electrons go
into the lowest energy orbits ($n=1$) and couple to
zero angular momentum; their combined wavefunction
is therefore spherically symmetric. The next eight
electrons go into orbits with $n=2$, which are less
tightly bound. They are therefore on average much
further from the nucleus than the other two, and in
addition six of the eight have one unit each of angular momentum which tends to keep them away from the
centre. Thus these electrons rarely penetrate the
spherical cloud which represents the other two and so
move around a central nucleus of effective charge
($Z-2$). This electron group corresponds to a distribution of charge in the form of a broad spherical
shell (in the classical sense), overlapping very
little with the shell below. So we build up a complex
atom shell upon shell, rather like an onion. This
picture is oversimplified and leads to incorrect numbers of electrons in shells ($2n^2$). The reason is that
the interpenetration of shells does take place. Because of the centrifugal effect sub-shells of lower
l will penetrate the shell below to a greater extent
than those of higher l (and the same value of n), so
removing the degeneracy enough to move sub-shells
effectively from one major energy shell to another.
A complete sub-shell, however, is spherically symmetric, so the overall picture is much the same but
the numbers of electrons in major shells is now different. This has a profound effect on the chemical
properties of the elements.

    In trying to adapt this approach to the nucleus,

a number of difficulties arise. In both cases the
potential well for the next particle must be built
up from the interactions of particles already present.
In the atom there is initially a strong attraction
to the centre, which carries virtually all the mass
and provides a basis for building up layers, in addition ensuring that the layers are physically well-
separated. A central force is therefore dominant at
all stages, since a uniformly charged spherical shell
behaves, in the region outside it, as though the
charge is concentrated at its centre. From Chapter 1,
however, it is known that there is nothing special
about the centre of the nucleus except that it is in
a region where the density is constant and where, from
symmetry, the average force on a nucleon is zero.
Thus the simplest form of potential which might approximate to the average nucleon-nucleus potential is
a square well, whilst the next approximation is to
make the potential well follow the shape of the density (see Fig.3.1) but extending further out by 1 fm

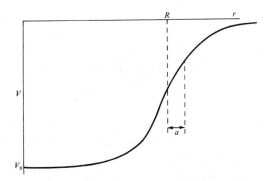

Fig.3.1. Saxon-Woods (or Fermi) approximation to the nucleon-
          nucleus potential.

or so, to account for the range of the nucleon-nucleon
force. With such a well the wavefunctions for the
single-particle eigenstates overlap strongly. The
centrifugal potential pushes the wavefunction away
from the centre for $l \neq 0$, but the effect is not mar-
ked for $l < KR$, where K is the wavenumber of the
nucleon inside the nucleus of radius R, and $KR \sim 10$
for a medium-weight nucleus.

In addition to greater mixing of wavefunctions
enhancing the importance of collisions, there is also
the fact that the nucleon-nucleon interaction is short-
range so that its contribution to the average potential is much less than its contribution to collision.
It would seem improbable that the picture of single-
particle orbits in an average central potential should
approximate to reality. But so far no account has
been taken of the Pauli principle. If one particle
gains energy in a collision, then the other particle
must lose it. In the ground state of a nucleus all
the lowest-lying single-particle states will be occupied, so no particle can lose energy. The same
two states must be occupied before and after the collision, so effectively there has been no collision.
The mean free path of a nucleon in the ground state
of a nucleus is effectively infinite, and the concept
of particle orbits is meaningful.

Having established the picture of nucleons in
well-defined states it is now appropriate to consider
whether these states group into shells, characterized
by having similar energies but not necessarily similar
spatial configurations as in atoms. In the case of
atoms, the periodic table preceded its explanation
in terms of the shell model, so for nuclei statistical

analyses of the properties of nuclear ground states revealed the 'magic numbers' which are the nuclear equivalent of the location of noble gases in the periodic table. We take a nuclear property such as binding energy B and plot it against A, or, better still, make a three-dimensional plot against N and Z. Thus in Fig.1.2 (p.19) the measured values of B/A differ from the smooth curve predicted by the mass equation in the direction of increased binding in the region of certain values of N or Z. Statistical analysis reveals the following facts:

1. A plot of binding energy of the last nucleon against N (or Z) reveals discontinuities at certain values of N and Z. (The neutron binding energy may conveniently be determined from the γ-spectrum following capture of thermal neutrons.)

2. Similar discontinuities in the binding of α-particles occur at $N = 82$ and 126.

3. Certain nuclei appear to be naturally more abundant than their neighbours in the plot of stable species. Certain elements have more stable isotopes than their neighbours, i.e. a greater variation of N for a given Z, and correspondingly there are values of N for which a greater variation of Z is permitted.

4. The excitation energies of the first excited states (2^+) of even-even nuclei are at maxima for certain values of N or Z, falling to broad minima between these values. In the broad minima the lowest states $(0^+, 2^+, 4^+)$ form a sequence in which the energy spacings $(0,2)$ and $(2,4)$ are in the ratio $6:14$; near the maxima these states

tend to be equally spaced.

5. Quadrupole moments of odd-*A* nuclei are large when
    neighbouring even-even nuclei have low-lying
    first excited states and are near zero when those
    states lie high.

6. Nuclear reaction cross-sections reveal periodic
    properties, e.g. thermal-neutron capture cross-
    sections (as shown in Fig.3.2).

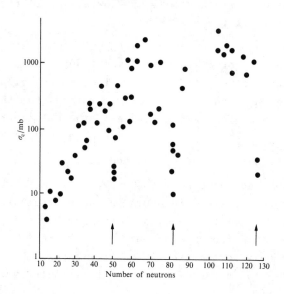

Fig.3.2. Effective capture cross-sections for reactor neutrons
         plotted against neutron number of target nucleus.
         Detailed explanation of this curve is complicated,
         but the influence of shell closure is apparent. (Based
         on CODD *et al.* (1956). *Prog.nucl.energy* 1, 296.)

7. The parities of nuclear ground states of odd-*A*

nuclei exhibit a systematic behaviour correlated
with the above 'magic' effects.

The nett result of such analyses is that nuclei
containing 2, 8, 20, (28), 50, 82, 126 neutrons or
protons (apart from 126) are particularly stable
(cf. the stability of noble gases). By analogy, shell
closure at these neutron or proton numbers is therefore expected. The single-particle energy levels of
nuclei should reflect such closure by a larger than
usual energy spacing at appropriate places in the
level diagram. Simply altering the shape of the central potential through permitted variations is unable
to produce energy gaps at these numbers (see Fig.3.3),
where the ordering of levels for four types of well
is shown to be unsuitable but the addition to the
potential of a spin-orbit term ($\equiv V(r)\underline{l}.\underline{s}$ for a given
nucleon) achieves success. This suggestion was made
independently by Mayer and by Haxel, Jensen, and Suess
in 1949. Such a term splits the otherwise degenerate
sub-shells $j = l + \frac{1}{2}$ and $j = l - \frac{1}{2}$ obtained by combin-
ing the spin and orbital angular momenta of a nucleon.
Since

$$\underline{s}.\underline{l} = \frac{1}{2}(j^2 - l^2 - s^2)$$

then

$$\langle \underline{s}.\underline{l} \rangle = \frac{1}{2}\{j(j+1) - l(l+1) - s(s+1)\},$$

giving the value $+\frac{1}{2}l$ for $j = l + \frac{1}{2}$ and $-\frac{1}{2}(l+1)$ for
$j = l - \frac{1}{2}$, except for $l = 0$, when both terms are zero.
Thus the splitting $\propto (2l+1)$; its increase with l is
important for the success of the modification since

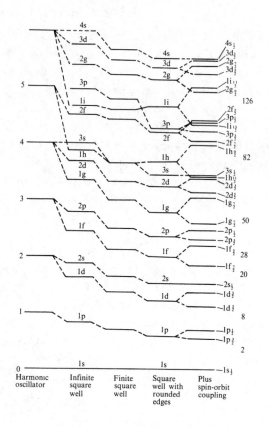

Fig.3.3. Locations of energy levels in four types of well, fol-
        lowed by the effect of a spin-orbit term on the fourth
        type. Note that the spin-orbit term is necessary to
        locate the magic numbers, as shown. (Based on FELD
        (1953). *A.Rev.nucl.Sci.* 2, 239.)

it serves to throw down, from one shell to another, one
of the components of the sub-shell with highest l.
By making the sign of the spin-orbit coupling opposite

to that arising from a magnetic term (as in atoms) the sub-shell corresponding to $j = (l+\frac{1}{2})$ is correctly thrown down. The order of the filling of neutron shells is then given by the sequence

$$(1s_{\frac{1}{2}})^2 \Big|_2 (1p_{\frac{3}{2}})^4 (1p_{\frac{1}{2}})^2 \Big|_8 (1d_{\frac{5}{2}})^6 (2s_{\frac{1}{2}})^2 (1d_{\frac{3}{2}})^4 \Big|_{20} (1f_{\frac{7}{2}})^8 \Big|_{28} (2p_{\frac{3}{2}})^4$$

$$(1f_{\frac{5}{2}})^6 (2p_{\frac{1}{2}})^2 (1g_{\frac{9}{2}})^{10} \Big|_{50} (2d_{\frac{5}{2}})^6 (1g_{\frac{7}{2}})^8 (3s_{\frac{1}{2}})^2 (2d_{\frac{3}{2}})^4 (1h_{\frac{11}{2}})^{12} \Big|_{82} (2f_{\frac{7}{2}})^8$$

$$(1h_{\frac{9}{2}})^{10} (3p_{\frac{3}{2}})^4 (2f_{\frac{5}{2}})^6 (3p_{\frac{1}{2}})^2 (1i_{\frac{13}{2}})^{14} \Big|_{126} (2g_{\frac{9}{2}})^{10} (3d_{\frac{5}{2}})^6 (1i_{\frac{11}{2}})^{12} (2g_{\frac{7}{2}})^8$$

$$\cdot \ \cdot \ \cdot \ \cdot \ \cdot \ \cdot \ \cdot$$

This sequence can be varied slightly by small changes in the spin-orbit potential without affecting the magic numbers.

The notation is slightly different from the atomic convention. The l_j convention is the same but the number preceding does not have the significance of n in the hydrogen-like spectra; it merely numbers the states of given l_j in the order in which they occur (it can be connected with the number of nodes in the radial wavefunction). The superscript is the number of nucleons which the sub-shell can accommodate $(2j+1)$. For protons, because of the influence of the Coulomb potential, the ordering is slightly different; it is the same up to 50 when the next shell looks like

$$(1g_{\frac{7}{2}})^8 (2d_{\frac{5}{2}})^6 (1h_{\frac{11}{2}})^{12} (2d_{\frac{3}{2}})^4 (3s_{\frac{1}{2}})^2$$

and above 82 the sequence is

$$(1h_{\frac{9}{2}})^{10} (2f_{\frac{7}{2}})^8 (3p_{\frac{1}{2}})^2 \ldots \ldots$$

These sequences, together with the pairing term, which is taken to couple pairs of like nucleons to zero angular momentum whenever possible, account for all the parities of the ground states of stable and nearly stable nuclei and, with a few exceptions, their spins for all even-A nuclei and up to $A \sim 100$ for odd-A nuclei. Two notable exceptions are ^{19}F and ^{23}Na which are $\frac{1}{2}^+$ and $\frac{3}{2}^+$ whereas the sequence of protons predicts $d_{\frac{5}{2}}$, i.e. $\frac{5}{2}^+$ for both. However, the $s_{\frac{1}{2}}$, $d_{\frac{3}{2}}$-states lie in the same shell and it was also discovered that a $\frac{5}{2}^+$ excited state lies low in each nucleus, so it was felt that 'residual forces' (i.e. forces left over after accounting for the well by averaging the nucleon-nucleon interaction) were moving these sub-shells relative to each other. Lately it is considered that nuclear deformation is responsible for these discrepancies.

Above $A \sim 100$ the ground states of odd nuclei do not appear to reflect the order in which the shells should fill; e.g. four isotopes of Sn ($Z = 50$) have ground states $\frac{1}{2}^+$ instead of only one as predicted. The reason lies in the pairing term; adding a second neutron into one of the isotopes $\frac{1}{2}^+$ should produce the pair $(3s_{\frac{1}{2}})^2$, but the pairing of neutrons in a higher lying state with higher l, probably the $1h_{\frac{11}{2}}$-state, could give an increased binding which more than off-sets the higher energy of the intrinsic state. In which case the pair of neutrons will be found in the $1h_{\frac{11}{2}}$-state in the next even-even isotope, with the $3s_{\frac{1}{2}}$-state unoccupied. So the next odd isotope of Sn will again be $\frac{1}{2}^+$, etc.

Also, just below this region, ground states $\frac{7}{2}^+$ should be found corresponding to neutron odd numbers 57 - 63. None is found, but in each nucleus where expected, a low-lying $\frac{7}{2}^+$ state occurs. A similar explanation can probably be made, though it is possible that the ordering of the $1g_{\frac{7}{2}}$- and $3s_{\frac{1}{2}}$-states should be inverted. For the more massive nuclei, nuclear deformation is expected to occur away from the closed shells, and this would produce a different ordering of the ground-state spins (see next section).

A triumph of the shell model concerns the location of 'islands of isomerism'. It had been noticed that long-lived isomers (excited states of nuclei which decay by γ-emission or its equivalent, internal conversion, with lifetimes long enough to be classified as radioactive) tended to cluster at certain regions of the periodic table. To understand this it is necessary to know something about γ-ray selection rules (see Chapter 5). At this stage we merely need to know that the bigger the spin change the slower the transition. In certain regions of the periodic table, the first excited state is expected to differ in spin from the ground state of an odd-A nucleus by

a large amount. For example, at $Z = 39$ the last proton is in a $2p_{\frac{1}{2}}$-state, whereas the first proton excitation is $1g_{\frac{9}{2}}$; it lies at $0 \cdot 381$ MeV in ^{87}Y, at
$0 \cdot 91$ MeV in ^{89}Y, and at $0 \cdot 551$ MeV in ^{91}Y, giving
lifetimes of 14 hours, 16 s, and 50 min respectively.
Thus it is seen that the spin-orbit term which produces the right sequence of shells also puts sub-shells
of high spin in juxtaposition with sub-shells of low
spin and so is responsible for isomerism.

Some mention must be made of the magnetic moments
of the odd-A nuclei. From the assumption of maximum
pairing, the properties of the nucleus should be those
of the odd nucleon. The magnetic moment contributed
by a single nucleon is

$$\mu = g_l \langle \underline{l} \cdot \underline{j} \rangle + g_s \langle \underline{s} \cdot \underline{j} \rangle \; \frac{m_{max}}{j(j+1)} \text{ with } m_{max} = j,$$

where $2 \langle \underline{l} \cdot \underline{j} \rangle = j(j+1) + l(l+1) - s(s+1)$, and correspondingly for $\langle \underline{s} \cdot \underline{j} \rangle$. This can readily be derived
using the vector model (see e.g. Kuhn 1961). By and
large the agreement with experimental data is poor.
It is best when the odd nucleon immediately precedes
or follows shell closure; examples for the odd proton
are $^{3}_{1}$H, $^{15}_{7}$N, $^{39,41}_{19}$K, $^{89}_{39}$Y, and for the odd neutron
$^{3}_{2}$He, $^{17}_{8}$O, $^{207}_{82}$Pb, but even here $^{41}_{20}$Ca, $^{45}_{21}$Sc, $^{91}_{40}$Zr show
large deviations. $^{19}_{9}$F might have been quoted above
were it not for the fact that theoretical computation
has come up with an untidy, three-particle mixture of
$s_{\frac{1}{2}}$, $d_{\frac{5}{2}}$, and $d_{\frac{3}{2}}$ wavefunctions outside an ^{16}O core,
that happens to give the magnetic moment of an $s_{\frac{1}{2}}$ pro-

ton. The deviations of magnetic moments are as yet
unexplained; they could arise from three sources: (1)
strict pairing does not occur, (2) nuclei may not be
spherical, and (3) the pion field around a nucleon inside a nucleus should differ from that around a nucleon
in isolation; this could give rise to a magnetic moment
differing from that obtained by vector addition of
the individual moments as ascribed to isolated nucleons.

Finally, how does the shell model cope with excited states? The general ordering of levels indicated
in Fig.3.3 should represent filled levels up to a certain value and single particle excitations beyond this.
If strict pairing does occur, the excited states of
an odd-A nucleus should follow this sequence of excitation. Such single-particle excitations do occur,
especially when they refer to large l-values in regions
of sub-shells of low l. However, the number of excited states up to, say, 3 MeV above ground is usually
far greater than the number of single-particle levels
available. We are forced to conclude that strict
pairing has broken down (it may well take only
$\sim$ 1 MeV to break a pair) and that extra states are obtained by the coupling of three or more particles. An
extreme example of this is ^{19}F. Its first excited
state, at 108 keV, is $\frac{1}{2}^{-}$ whilst the lowest single-
particle states of odd parity $(1f_{\frac{7}{2}})$, $(2p_{\frac{3}{2}})$ should be
several MeV higher. Merely decoupling the last two
neutrons will not produce odd parity since all three
particles are in even-parity orbits - the $s_{\frac{1}{2}}$, $d_{\frac{5}{2}}$, and
$d_{\frac{3}{2}}$ which lie close together. It has therefore been
assumed that the closed proton p-shell of the ^{16}O core

has been disrupted, and one of the protons has gone
into the s-d shell to create something like an α-
particle; the $(p_\frac{1}{2})$ 'hole' provides the odd parity for
this state and others at somewhat higher excitation.
Thus, where excited states are concerned, even the
closed shells must be considered vulnerable.

The shell model forms a working basis for attempting to determine the level scheme of a nucleus, but
there are interactions between the nucleons, termed
'residual forces' since they remain after accounting.
for the well by averaging out the main body of the
forces. These residual forces are often arrived at
empirically in building up nuclei progressively from
a double-closed shell core by comparison with experimentally determined level schemes.

COLLECTIVE MOTION

In this and the previous chapters, reference has
been made to a collective nuclear motion. For this
to occur, motion of closed shells is implied, and this
in turn means that such shells cannot be spherically
symmetric - quantum mechanically there can be no motion about an axis if the system has symmetry about
that axis. Thus the occurrence of collective motion
is characterized by a departure from spherical symmetry of the nuclear system; it may be the rotational
motion of a permanently deformed nucleus or the motion
of a nucleus vibrating (possibly about an undeformed
mean structure) through different degrees of deformation.

The possibility of permanent deformation was investigated by Nilsson, who calculated the single-par-

ticle states in a spheroidal potential well having a
symmetry axis. The distortion of the spherical well
is then characterized by the single parameter $\delta = \Delta R/R$,
as defined in Chapter 1. For such a well *fixed in
space* it must be appreciated that, because of the absence of spherical symmetry, the total angular momentum J is no longer a constant of the motion, but because the system does have symmetry about one of its
principal axes the component of angular momentum along
this axis, designated by K, is indeed a constant of
the motion. The behaviour with δ of the low single-
particle levels is shown in Fig.3.4, where the value
of $|K|$ is shown on each curve. Since the deformation
is quadrupole no distinction is made between up and
down, so that each curve has degeneracy 2 corresponding
to $\pm K$. In the limit $\delta = 0$, the spherical states
appear; in particular the $d_{\frac{5}{2}}$-state appears as the con-
fluence of three states of different $|K|$ and there-
fore has degeneracy 6, which is in accord with its
spatial degeneracy, $2j + 1$.

   If the problem is made physically more realistic
by isolating the well in space rather than fixing it,
then the well itself will be capable of rotation, but
the rotation must have no component along the symmetry
axis. If the rotational motion is slow compared to
the single-particle motion (the adiabatic limit), then
the latter will have a well-defined angular-momentum
component along the nuclear axis K, but j itself will
not be well-defined. The total angular momentum J will
also be well-defined (an isolated body) and will have
a component K along the nuclear axis and a component
M along the external axis of quantization. $\underline{J}$ will be

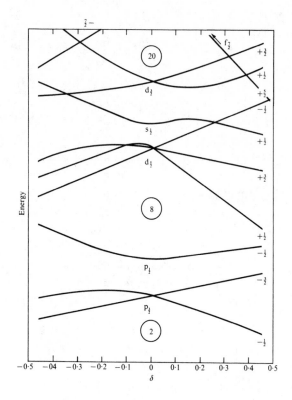

Fig.3.4. Single-particle states in a fixed ellipsoidal potential
well, characterized by $\delta = \Delta R/R$. The states are
labelled with $|K|$ and parity (they are doubly degenerate)
when $\delta \neq 0$. At $\delta = 0$ they correspond to the sub-states
of the usual shell-model states. The $^1S_{\frac{1}{2}}$-state has
been omitted. (Based on NILSSON (1955).[2] *Dan.mat.
-fys.medd.* 29, No.16.)

a vector sum of $\underline{j}$ and $\underline{R}$, the nuclear rotation (see
Fig.3.5). The rotational energy will be given by

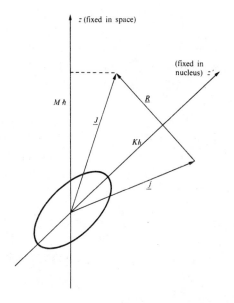

Fig.3.5. Coupling scheme for a deformed nucleus. $\underline{J} = \underline{j} + \underline{R}$, $j_{z'} = J_{z'} = K\hbar$, $J_z = M\hbar$. Note that $\underline{j}$ itself is not a constant of motion, nor is j^2 well defined; e.g. $J^\pi = \frac{7}{2}^-$ can be constructed from $f_{\frac{7}{2}}$, $h_{\frac{9}{2}}$, $h_{\frac{11}{2}}$ orbitals coupled to $R = 0$ or 2, and, except for extremes of coupling, all components will be present.

$$E_R = (\hbar^2/2I)\langle R^2 \rangle = (\hbar^2/2I)\langle (\underline{J}-\underline{j})^2 \rangle$$

$$= (\hbar^2/2I)(J(J+1) + \langle j^2 \rangle - 2\langle \underline{J}\cdot\underline{j} \rangle)$$

$$= (\hbar^2/2I)(J(J+1) + \langle j^2 \rangle - 2K^2), \qquad (3.1)$$

since it can be shown that

$$\langle J_x j_x + J_y j_y \rangle = 0,$$

unless there are components of $K \pm 1$ as well as K in
the wavefunction. Since there are components of $\pm K$
due to up-down degeneracy, these terms are not zero
for $|K| = \frac{1}{2}$, but this case will be ignored. The term
in $\langle j^2 \rangle$ can be incorparated into the other single-
particle contributions to the energy, and forgotten.

This expression defines a set of rotational states
based upon a common wavefunction relative to the nuclear axis. The lowest-lying state will have $J = K$
(the lowest value J can have for a component K along
the nuclear axis) and a degeneracy $2K+1$. So, for the
isolated nucleus, the confluence at $\delta = 0$ for the
$j \equiv d_{\frac{5}{2}}$ configuration will no longer have the correct
number of sub-states and would appear to have three
values of $J(\frac{1}{2}, \frac{3}{2}, \frac{5}{2})$ instead of the $\frac{5}{2}$ state of the
spherical nucleus.

The reason for the discrepancy is that the rotational energy has been ignored in Fig.3.4. How to
evaluate this depends upon the nature of the rotation.
It could be a rigid-body rotation, or an effective
rotation of a hydrodynamical fluid, an example of
which is the tidal deformation of the earth by the moon
On the surface of the earth this appears like a rotation, but the motion is in fact up and down. Experimental investigation indicates that the latter is
nearer the truth and $I \sim I_0 \delta^2$, where I_0 is the rigid-
body moment of inertia of a sphere, namely, $\frac{2}{5} MR^2$.
The rotational energy therefore becomes very large at
small δ so the states $J = \frac{1}{2}$ and $\frac{3}{2}$ produced from a $d_{\frac{5}{2}}$
particle move up to very high energies as δ becomes
small. More correctly, the coupling scheme breaks
down; it is no longer valid to couple the single-par-

ticle motions rigidly to the well and then to allow
the whole system to rotate. The figure only approxi-
mates to reality when δ is sufficiently large for the
rotational energy to be small compared to the spacing
of the different particle configurations.

At sufficiently large δ, the diagram correctly
predicts the ordering of states. The particles couple
strongly in pairs with equal and opposite K, so an
even-even nucleus will always have $K = 0$ and a ground
state $J = 0$. Putting $J = 0$, $K = 0$ in the rotational
energy gives a rotational band of levels of energies,
relative to the ground state, $J(J+1)\hbar^2/2I$, where because of symmetry requirements (up $\equiv$ down) only even
values of J can occur. This symmetry requirement is
met by expressing the wavefunction of the system as
a combination of the wavefunctions $(J,M,+K)$ and
$(J,M,-K)$ thus: $(\psi^J_{M,+K} + (-1)^{J-K} \psi^J_{M,-K})$. For $K = 0$
it is seen that the two components are the same except
for the term $(-1)^J$ which eliminates all odd J. For
other values of K no such restriction arises, so J
increases in unit steps up a band. For odd-A nuclei,
the sequence of levels in Fig.3.4 determines a value
of $|K|$ for the ground state. A rotational band is
now built up having energies relative to ground
$\hbar^2\{J(J+1)-K(K+1)\}/2I$. It should be noted that $K=\frac{1}{2}$ produces a more complicated ordering of states which will
not be discussed, except to mention that it arises from
the $\langle \underline{J}.\underline{j} \rangle$ term in the rotational energy.

A typical level scheme is shown for an even and
an odd nucleus in Fig.3.6. It will be seen that rotational states up to large values of J are obtained
and that they tend to be lower lying than the next
particle configuration. Their locations usually con-

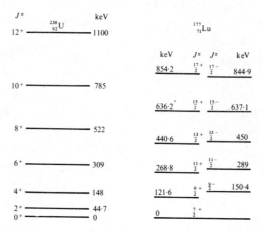

Fig.3.6. Examples of rotational bands in even-*A* and odd-*A*
         nuclei. In the latter nucleus the level schemes
         representing the two non-interacting bands have been
         separated laterally. For clarity some levels un-
         connected with the bands have been omitted from both
         schemes.

form well with the simple expressions above, but there
is usually a small discrepancy which increases with *J*
and which is ascribed to an increase of I due to a centrifugal stretching of the nucleus. Experimentally,
levels in a rotational band are revealed by the very
large E2 transitions (see Chapter 5) between members
- the intrinsic strength can be as large as 200 times
single-particle strength in heavy, deformed nuclei
in the rare-earth region. Such large enhancements
arise from the rotation of the nucleus as a whole,
corresponding to a correlated motion of up to *Z* protons and therefore capable in principle of producing

enhancements up to z^2.

Finally, nuclei undeformed in the ground state can have excited states corresponding to vibrations through varying degrees of deformation. As in the case of the harmonic oscillator, the different vibrational states are equally spaced, but because the deformation is quadrupole, the vibrational quanta are characterized by 2^+ giving for an even-even nucleus the sequence of states (0^+), (2^+), $(0^+,2^+,4^+)$, $(0^+,2^+,3^+,4^+,6^+)$, etc. where the brackets correspond to 0, 1, 2, 3, etc. vibrational quanta coupled using

Fig.3.7. A vibrational system. The next levels are located around 1·6 MeV and could include members of the three-quantum vibrational group of states.

Bose statistics. The bracketed states are degenerate
in the model, but departure from the model removes the
degeneracy as Fig.3.7 shows.

THE OPTICAL MODEL

The previous models refer to a static or quasi-
static system since the excited states so far introduced are bound states decaying only by the relatively weak radiation process and therefore having lifetimes much longer than a characteristic nuclear time,
which is the time taken for a nucleon to traverse
a nuclear diameter. In this situation the quantum-
mechanical problem consists of setting up stationary
states (just as for the ground state) and then treating
the radiative process as a perturbation. When the
energy of the system is large enough to permit nucleon
(or composite particle) emission then the decay is
via the strong interaction so the state can have a
lifetime of the order of the characteristic time.
The decay process is no longer a small perturbation
and a different approach is needed. The optical model
meets this need.

This model attempts to cope with the quantum-
mechanical problem of the scattering and absorption
of particles impinging on a nucleus. Since a simple
potential has been used to represent the nucleus in
the shell model, it is appealing to do the same for
the scattering problem. An argument for the use of
such a potential was based upon the long mean free
path of a nucleon inside the nucleus as a consequence
of the Pauli principle. However, a nucleon coming in
from outside has $(E_{kin} + 7)$ MeV (the incoming kinetic
energy + binding energy) more energy than any nucleon

within the nucleus. It can collide with such a nucleon thereby losing energy and exciting that nucleon.
Because of the Pauli principle it cannot interact with
nucleons lying more than $(E_{kin} + 7)$ MeV below the
Fermi surface of the nucleus. Thus as the kinetic
energy of the incident nucleon increases, the limita-
tions of the Pauli principle are reduced and the mean
free path of this nucleon gets shorter.

How is this concept, fed into the problem? A beam
of particles traversing a region in which the mean
free path of a particle is λ will decrease in beam
intensity with distance as $I = I_0 \exp(-x/\lambda)$. If the
beam of particles, without absorption, is represented
by the plane wave amplitude $\sqrt{I_0} \exp(ikx)$, then with
absorption it should be

$$\sqrt{I_0} \exp(-x/2\lambda)\exp(ikx) = \sqrt{I_0} \exp\{i(k+i/2\lambda)x\}.$$

The wavenumber has therefore been modified from k to
$(k + i/2\lambda)$. But $\hbar k = \{2m(E-V)\}^{\frac{1}{2}}$, where E is the total
energy of the particle and V the potential energy.
Putting in the modified expression gives

$$k^2 - (1/2\lambda)^2 + ik/\lambda = 2m(E-V').$$

Since the total energy of the particle must be real
(it can move outside the region of absorption to be
measured), an imaginary component must be added to
the potential. Expressing this as $V+iW$ gives
$2m(E-V)/\hbar^2 = k^2-(1/2\lambda)^2 \approx k^2$ for reasonably large λ;
and $2mW/\hbar^2 = -k/\lambda$ connects the imaginary component
with the mean free path.

In its early stages therefore the optical model

used a complex square-well potential, $-(V+iW)$ for
$r < R$, zero for $r > R$. This was later changed to the
Saxon-Woods potential

$$\frac{-(V+iW)}{1+\exp\{(r-R)/c\}} ,$$

where V, W, R, and c were looked upon as arbitrary
parameters in fitting the experimental scattering data
at a given energy. The latter two parameters should
obviously correspond in some way to the density parameters; the simplest way being to assume that the
potential follows the density when allowance is made
for the range of nuclear forces. The real term V
might be expected to approach the shell-model potential
at low enough energy. There is no good reason why V
and W should behave in the same way; indeed, an energy-
dependence of W is expected from the previous argument
which is unlikely to be similar to that of V. Also
it can be argued that the bound nucleons near the nuclear surface should be those with greatest energy,
for which collisions with the incoming nucleon are
least restricted. Thus a peaking of W near the surface may be more realistic, and this can be achieved
by associating W with the derivative of the Fermi
(Saxon-Woods) function, or by simply using a Gaussian
function peaked at the surface. In addition, for
treatment of polarization data, it has been found
necessary to introduce a spin-orbit coupling term,
$\propto \underline{l}.\underline{s}$, where $\underline{l}$ and $\underline{s}$ refer to the incident nucleon.

It is instructive to consider s-wave scattering
from a square-well potential of radius R in the limit
of very small W. In order to avoid the complexities
of determining elastic scattering cross-sections, the

variation with energy of the reaction cross-section
is estimated; for $W=0$, there is no absorption and
therefore only elastic scattering, but for W small
enough the simplifying features of $W=0$ can be used.
The wavefunction inside the well (for $W=0$) is of the
form $(A/r)\sin Kr$, satisfying the requirement of a node
at the origin in r times the radial wavefunction,
whilst outside the well the solution can be written
$(B/r)\sin(kr+\delta)$, where K and k are the interior and
exterior wavenumbers: $K = 2m(E+V)^{\frac{1}{2}}/\hbar$ and $k = (2mE)^{\frac{1}{2}}/\hbar$.
The standing-wave solutions indicate that no absorption
has taken place, i.e. as many neutrons are moving away
from the nucleus as toward it (and the flux of neutrons
toward the nucleus $\propto k|B|^2$, though the direct con-
nection is complicated since the neutrons proceeding
toward the nucleus constitute a plane wave).

The continuity of the wavefunction and its derivative at the nuclear surface lead to the equations

$$k\tan KR = K\tan(kR+\delta)$$

and

$$A = Bk/(K^2\cos^2 KR + k^2\sin^2 KR)^{1/2}.$$

If for simplicity it is assumed that $K \gg k$ then A/B
is seen to vary between maxima of ~ 1 when $KR \sim (n+\frac{1}{2})\pi$
and minima of $\sim k/K$ when $KR \sim n\pi$. Also $|A|^2$ will fall
to half its maximum value when $KR \sim (n+\frac{1}{2})\pi \pm k/K$.

If now a very small imaginary term is introduced
into the potential, then absorption will take place
within the nucleus, but for small enough W this will
hardly change the wavefunctions. From the connection

between W and mean free path it is obvious that the re-
action yield will be $\propto W|A|^2$. The reaction cross-
section will therefore be $\propto W|A|^2/k|B|^2 = Wk/(K^2\cos^2 KR +$
$+k^2\sin^2 KR)$, showing maxima and minima with variation
of neutron energy. If, as in Chapter 4, the problem
were approached in a different way, namely, by inserting the neutron inside the nucleus and observing
its transition out, then it will be seen that the maxima correspond to the formation of 'virtual states' of
the system. The reaction is said to exhibit 'resonance'
when the neutrons are at an energy appropriate to one
of the virtual states within the nucleus, cf. forced
oscillations of a pendulum near its natural frequency
or of an electric circuit. The ratio of resonance
half-width to resonance spacing can be estimated from
the above equations as $(2k/K) \times 1/\pi$ or $(4k/K) \times 1/2\pi$.
The ratio has been expressed in this latter way since
the factor $4k/K$ is the mismatch factor at the potential step (see Chapter 4 and Problem 4.1), whilst
the factor $1/2\pi$ represents an important theorem concerning widths of virtual states, namely, that on
average the decay width of a state into any one channel,
after allowing for matching the internal and external
wavefunctions, should be $1/2\pi$ times the local spacing
of levels with the same spin and parity. For a medium-
sized nucleus and a well-depth of 50 MeV, the level
spacing of s-wave single-particle neutron states
($J^\pi = \frac{1}{2}^+$ for $J^\pi = 0^+$ targets) is about 15 MeV and
the single-particle width about 2 MeV. Thus the single-
particle resonances are broad but well separated.

In going to a more realistic model of the nucleus,
as well as the single-particle states there are also
found vast numbers of levels of the same spins and pari-

ties but corresponding to excitations of more than one particle. In fact the single-particle states alone cannot produce reaction but only re-emission of the same particle. That reaction does take place indicates that the single-particle state is mixed in (by some unspecified perturbation) with all these more complicated states and W must in some way represent the degree of mixing. In the mixed-up system the theorem concerning level width and spacing will on average still hold (see below), though the mixing may reduce the spacing from a few MeV to a few eV.

Inclusion of a large imaginary term in the potential will broaden out the resonance even more (the absorption term is representing other channels into which the state can decay, so the width should roughly increase by the amount W (see Chapter 6)). Since absorption has taken place, followed by decay into another (reaction) channel, a broad resonance is to be expected in this reaction channel, e.g. the emission of inelastic neutrons. Data on total neutron cross-sections are presented in Fig.3.8 and also the results of analysis in terms of the optical model. A feature of the model is that the parameters which describe the potential should vary only slowly from nucleus to nucleus, and this appears to be the case. Typical values for the potential for neutron scattering from a wide range of nuclei at neutron energies up to $\sim$ 50 MeV are

$$V = 52 \cdot 5 - 0 \cdot 6 \, E_n \text{ (MeV)},$$

$$W = 2 \cdot 5 + 0 \cdot 3 \, E_n \text{ (MeV)},$$

$$R = r_0 A^{1/3}, \text{ where } r_0 = 1 \cdot 20 - 1 \cdot 25 \text{ fm},$$

$$c = 0 \cdot 5 - 0 \cdot 6 \text{ fm}$$

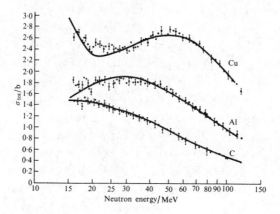

Fig.3.8. Plot of total neutron cross-section against energy for
C, Al and Cu. The full lines show optical-model fits.
(Based on BOWEN *et al.* (1961). *Nucl.Phys.* <u>22</u>, 640.)

The success of the model is illustrated better
by measurements of angular distributions as shown in
Fig.3.9. The situation is paralleled in optics by
diffraction through a circular aperture or, better
still, the scattering (diffractive) of a small transparent drop of liquid of refractive index different
from unity. Also, perhaps paradoxically, the model
has had a great deal of success in the analysis of
slow neutron data, where compound-nucleus (Chapter 6)
resonances of width typically measured in fractions
of 1 eV are observed. Representing the nucleon-
nucleus interaction by a simple central potential
cannot possibly account for these manifestations of
the detailed structure of the nucleus, but it has

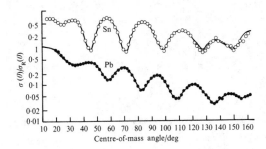

Fig.3.9. Differential cross-section (expressed as a ratio to
         Rutherford) for elastic scattering of 30·3 MeV protons,
         showing optical-model fits. (Based on SATCHLER (1967).
         Nucl.Phys. A92, 273.)

been shown that the optical model predicts average
cross-sections where the range of averaging covers
many compound nucleus resonances. For W very large
($\sim$ single-particle spacing) then the ratio of level
width (without barrier) to spacing, which determines
the average cross-section, will be of the order of
$1/2\pi$; but for W small this ratio, known as the
'strength function', will follow the shape of the
optical-model resonance.

PROBLEMS

3.1. The lowest states in $^{207}_{82}$Pb are (energy in MeV in
     brackets) $\frac{1}{2}^-$ (0), $\frac{5}{2}^-$ (0·57), $\frac{3}{2}^-$ (0·90), $\frac{13}{2}^+$ (1·63),
     and $\frac{7}{2}^-$ (2·34). Interpret these in terms of the
     shell model.

3.2*. $^{9}_{4}$Be, $^{13}_{6}$C, $^{17}_{8}$O have ground state magnetic moments
      - 1·17 nm, +0·07 nm, and -1·89 nm respectively.
      Deduce values for (a) the spins and parities,

(b) the magnetic moments of these nuclei, as predicted by the shell model. Comment on any discrepancies.

3.3*. Under what circumstances are (a) the single-particle shell model, (b) the rotational model expected to give a good description of the properties of nuclei. Describe two features of nuclear
behaviour for which either model makes characteristic predictions and show how these predictions can be tested experimentally.

3.4*. The energy levels of the $K=0$ ground-state rotational band of ^{238}U are given in Fig.3.6 (p.78).
Estimate the moment of inertia I by equating the
excitation energies with rotational energies.
Why does I increase with increasing angular
momentum? Why does the sequence contain only
even values of J? Compare I with $I_{rigid} = \frac{2}{5} AMR^2$
for the appropriate rigid-sphere rotation and
comment on the difference.

3.5. Use Bose statistics to show that three 2^+ vibrational quanta can couple to form the states
$J^\pi = 0^+$, 2^+, 3^+, 4^+, 6^+.

3.6. The optical model potential for neutrons on a
medium-weight nucleus ($A \sim 125$) is given by the
combinations $(0,50,2), (30,40,8), (100,25,8)$, and
$(300,10,5)$, where the first number in a bracket
is the neutron energy, the second $-V_0$, and the
third $-W_0$ all in MeV). Determine the mean free
path at each energy near the centre of the nucleus; show that the nucleus is reasonably trans-

parent at each end of the range, and even at maximum absorption there is $\sim$ 10 per cent chance of a neutron passing diametrically through the nucleus. Explain the general trend of V and the mean free path. If the nucleus is transparent for near-thermal neutrons, how is it that they can be captured for long periods of time (see Chapter 6)?

3.7. Work through the problem outlined on p.83 and verify the two equations given there.

4. Spontaneous decay of Nuclei I: α and ß decay

In Chapter 1 it has been shown that the stable nuclei
are surrounded by neighbours which are unstable to
ß-decay, whilst further away from the stability line
nucleon emission can take place. These latter nuclei
can only be observed in nuclear reactions. For mass
values greater than 150, even the nuclei on the stability line become unstable, not to nucleon emission
but to emission of α-particles because the binding
energy per nucleon of the α-particle is similar to
that of a nucleus.

α-DECAY

Historically, α-particle decay was observed in
the heavier nuclei ($A > 206$), and the lifetime for
decay was seen to be very strongly correlated with the
energy of the α-particle emitted (because the process
is a break-up into two particles, the energy of the
α-particle is well defined for unique parent and
daughter states). For these nuclei all observed α-
emissions lie within the energy limits
4 MeV $< E_\alpha <$ 9 MeV, the lower value being associated
with a lifetime $\sim 10^{+18}$ s and the upper value with
10^{-6} s, according to the Geiger-Nuttall rule which can
be expressed as $\ln \lambda = a + b \ln E$, where λ^{-1} is the
mean life. This rule explains why, although almost
all nuclei in the range A = 150 - 200 are unstable
to α-emission, only a few have been observed to decay
in this way; in general the available energy is very
low and the lifetime too long to permit detection of
the process. An example of an observed decay in this

region is $^{152}_{64}$ Gd, with a lifetime of 10^{14} years and an α-particle energy 2·24 MeV. These figures are not compatible with the Geiger-Nuttall relationship as applied to the natural radioactive series, but it will be seen later that the connection between lifetime and energy depends very strongly on the nuclear charge Z.

Another historic experiment was the bombardment of the daughter nucleus with α-particles of the same energy as (or even greater than) those emitted by the parent. The scattering was observed to be of the pure Rutherford type, which, on the face of it, indicated that the nucleus was still behaving effectively as a point in its interaction with an α-particle approaching it from outside, even though this particle was closer to it than the region from whence (in the classical sense) an α-particle must have originated in the decay of the parent. This paradox was explained by Gamow in his theory of barrier penetration.

In Fig.4.1 an idealized form is presented of the

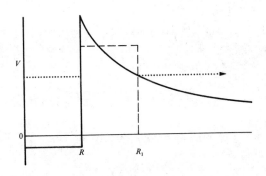

Fig.4.1. Idealized potential for s-wave α-particle plus product nucleus. Also shown is the potential step approximation.

potential to which an incoming or outgoing α-particle
is exposed. The shape of the potential is dominated
by the Coulomb interaction at long distances and by
nuclear forces at short distance. These latter have
been simplified to a square-well form, though they
should conform to something like the optical potential,
which will smooth out the sharp discontinuity and also,
by means of the imaginary term, produce absorption
within the nucleus. Also shown in Fig.4.1 is a cruder
approximation to the potential in terms of potential
steps.

It is instructive to solve the wave-mechanical
problem, in one dimension, of a plane wave of wavenumber K impinging from the left onto this latter
potential, being partially transmitted, and finally
emerging as a wave of wavenumber k in the far right
region; in the intermediate region it must be represented by real exponentials (multiplied by complex amplitudes) if, in this region, its total energy is less
than the height of the potential step. This exercise
is left to the reader.

If in this problem the incident wave is of the
form $I\exp(iKr)$ and the transmitted wave $A\exp(ikr)$ then
the transmission coefficient T is defined by the ratio
of transmitted to incident flux

$$T = v_0|A|^2/v_I|I|^2 = k|A|^2/K|I|^2,$$

where v_I, v_0 are ingoing and outgoing velocities. For
this particular problem

$$T = \frac{16Kk\gamma^2}{(K^2+\gamma^2)(k^2+\gamma^2)} \exp(-2\gamma\Delta), \qquad (4.1)$$

where $\gamma = \{2m(V-E)\}^{1/2}/\hbar$ and Δ is the width of the
barrier, (R_1-R).

There is a direct correspondence between the
solution of a one-dimensional problem and the radial
dependence of the solution of the corresponding
three-dimensional problem with spherical symmetry.
The one-dimensional wavefunction $U(r)$ is simply connected to the radial dependence of the three-dimensional wavefunction $F(r)S(\theta,\phi)$ by $U(r) = rF(r)$, but
the potential for the one-dimensional problem is not
just $V(r)$ but $V(r) + l(l+1)\hbar^2/2mr^2$, i.e. allowance
must be made for the centrifugal potential. Thus the
solution above is appropriate for transmission of an
$l = 0$ α-particle through the spherically symmetric
potential step. Also the transmission coefficient
is unchanged since the flux at r is $v(r).r^2|F(r)|^2$ in
the three-dimensional case and therefore $\propto v|U|^2$ as
before. It has been assumed that $S(\theta,\phi)$ is correctly
normalized, i.e. $\int|S(\theta,\phi)|^2 d\Omega = 1$.

However, considered as a three-dimensional problem, there must be a node (in $U(r)$) at the origin.
In principle, no simple solution as obtained can give
such a node since, to conserve flux, the reflected
wave on the left (i.e. inside the spherical potential)
must have smaller intensity than the incident wave.
But the difference ($\sim \exp(-2\gamma\Delta)$) may be very slight,
and to good approximation can be ignored. The requirement of a node at the origin in addition to the
(amplitude and) phase relationship between the two
components derived from solution of the previous problem overdetermines the solution within the potential,
leading to a discrete set of solutions corresponding
to energy levels of the α-particle. These solutions

are not quite stationary states since they are decaying via transmission through the barrier.[†] However,
if $\exp(-2\gamma\Delta)$ is small then the decay is slow when measured on a nuclear time scale. The decay constant λ_α (in
s^{-1}) will be the outgoing flux when the wavefunction inside the 'nucleus' of radius R is normalized to unity:
this normalization is realized by putting

$$\int_0^R |I\exp(iKr) + J\exp(-iKr)|^2 dr = 1,$$

where $J\exp(-iKr)$ is the reflected wave. Because of
the node at the origin the integral becomes
$\int_0^R 4|I|^2\sin^2 Kr \, dr$ giving $|I|^2 \approx 1/2R$, since the point
R is very close to being an antinode. From the un-
certainty principle,

[†] A more accurate solution of the problem introduces difficulties of normalization. The outgoing wave
must slowly increase in amplitude as the distance
from the nucleus increases, representing the fact that,
at an earlier time, the probability of finding an α-
particle within the nucleus was greater, and therefore
the leakage through the barrier was greater. Thus
the normalization integral diverges outside the nucleus. The approximate solution given can be looked
upon as a perturbation approach in which an infinite
confining potential has been switched off at $t = 0$ and
the rate of increase of the final state (corresponding
to a free α-particle) is determined when the initial
state is known to represent one α-particle confined
within the nucleus.

$$\Gamma_\alpha = \hbar\lambda_\alpha = \hbar v_0 |A|^2 \quad \text{when} \quad |I|^2 = 1/2R,$$

$$\text{i.e.} \Gamma_\alpha = \frac{\hbar^2 k}{m} \cdot \frac{|A|^2 K |I|^2}{K|I|^2} = \frac{\hbar^2}{m} \cdot T \cdot \frac{K}{2R} = \frac{\hbar^2}{2mR^2} \; (KRT). \qquad (4.2)$$

This expression separates the α-decay width into two parts: the part $\hbar^2/2mR^2$, which has dimensions of energy (width) and is the nuclear scale factor for width, and the dimensionless bracketed term, which is the transmission coefficient suitably modified for an effectively bound α-particle rather than a flux. This transmission factor is the product of a major term, the barrier penetrability, and a minor term which here represents the product of two mismatch factors arising from the two discontinuities in the potential (cf. reflection of optical waves at a glass surface or vibrations at a joint between stretched strings). In a more realistic calculation, introducing the Coulomb potential, it is reasonable to expect (and more detailed calculations confirm this) the penetrability term to be replaced by $\exp\{-2 \int_R^{R_1} \gamma(r)\,dr\}$, since γ is now a varying function of r. The limits of integration R and R_1 are the points where $\gamma(r) = 0$. The slowly varying γ effectively changes adiabatically to an equally slowly varying k without discontinuity and so eliminates the mismatch factor at R_1. At R the discontinuity of the nuclear square-well potential will introduce a mismatch factor, but a (rapidly changing) Saxon-Woods-type potential will reduce the mismatch though not necessarily eliminate it. This accounts for the fact that analysis in terms of a square well appears to require a potential of approximately zero inside the nucleus whereas analysis in terms of the optical model usually gives a negative

internal potential (-25 MeV or so), though with considerable latitude since there are other variables in
the analysis. Note that the modified barrier penetrability may also be applied to the case $l \neq 0$, since
$\gamma(r)$ will be computed with the centrifugal barrier
included.

The chief source of uncertainty in the calculation of decay widths and lifetimes is in deciding
what fraction of the single particle width $\hbar^2/2mR^2$
must be used for real nuclei. The original calculations employed the full single-particle value, as
given here, which for nuclei with $A > 200$ is about
0.1 MeV, though the precise value will depend on
the nuclear potential adopted. It was later suggested
that 1 eV may be more realistic; this estimate was
based upon widths of resonances in slow neutron reactions (see Chapter 7). Such a large reduction in
nuclear factor was off-set in the barrier-penetrability
term by employing a larger nuclear radius. Nowadays,
these radii so deduced are considered to be too large,
which is to be expected since the levels from which
the width was inferred are at an excitation of order
7 MeV and statistical theories of nuclear levels require that as level densities increase then level
widths (for a particular channel, though not necessarily total widths) should diminish. Somewhere near 1
per cent of the single-particle width is now considered a reasonable estimate. Obviously, in face
of such uncertainties, variations in mismatch factors
due to details of the nuclear potential are of little
significance.

An examination of the α-spectrum of a given
radioactive nucleus reveals in general more than one

transition energy. The ground state of the parent
nucleus can decay to a number of states in the daughter
nucleus (of which the ground state may be one).   In
general too the higher the excitation of the daughter
state the smaller the intensity of the transition,
since the penetrability is so strongly dependent upon
the α-energy. For the naturally occurring α-emitters
a rough estimate is a factor of 3 decrease in transition probability for 100 keV decrease in α-energy.
In addition, the angular momentum requirements will
contribute to intensity variations. Table 4.1 gives
a rough guide to this effect.

TABLE 4.1

---

l	0	1	2	3	4	5	6	7	8
Reduction factor	1	0·7	0·37	0·14	0·04	7×10^{-3}	1×10^{-3}	1×10^{-4}	7×10^{-6}

As an example consider the decay $^{238}_{94}\mathrm{Pu} \rightarrow {}^{234}_{92}\mathrm{U} + \alpha$.
Table 4.2 gives the measured transitions to states of
known J (since $^{238}_{94}\mathrm{Pu}$ has $J = 0$ in the ground state,
the final J value is also the l-value for the emitted
α-particle).

The first five transitions are to members of a
rotational band. Such states have similar wavefunctions since they represent the same nuclear configuration in different degrees of rotation; it is therefore
gratifying to see that the decrease in transition
strength is largely accounted for by the variation of
penetrability with energy and angular momentum. The

TABLE 4.2

J	0^+	2^+	4^+	6^+	8^+	1^-
E(keV)	0	43	143	297	499	786
Relative strength:						
Estimated	1	0·25	9×10^{-3}	3×10^{-5}	3×10^{-8}	1×10^{-3}
Observed	1	0·39	$1\cdot2\times10^{-3}$	7×10^{-5}	1×10^{-8}	2×10^{-7}

sixth transition (others are given in the literature for this decay, but this one is typical) has a much weaker strength than would be accounted for by these means. It therefore reflects the nuclear contribution to the process. Thus the conclusion to be drawn is that for members of a rotational band (or for that matter a vibrational band) the simple estimates of relative strengths should be fairly good, but in general the dominant factor is the nuclear intrinsic width when low-lying states are being compared.

Finally, mention must be made of 'long-range' α-particles (this historical term is synonymous with 'high-energy'). An excited state can, in principle, emit α-particles; indeed its decay probability into the α-channel will probably be greater than that of the ground state. However, it will be competing un-favourably with γ-emission (see Chapter 5) at least for states near ground; at high enough excitation the reverse will occur but such states are not excited in the natural α-decay series. If the energy available to the α-particle is very large

then the decay probability will be correspondingly
large.   If also the γ-transition probability is
reduced because of high spin change, for example, then
it may be possible that the excited state can have
an observable α-decay. The key word is observable;
modern techniques have revealed quite large numbers
of 'long-range' transitions giving low-intensity α-
groups above an intense ground-state group. From what
has been said they are most likely to be found in conjunction with short-lifetime ground-state decays. An
example is $^{212}_{84}$Po, with a ground-state half-life of
304 ns and E_α = 8·79 MeV, having three such transitions, of which the most intense, $1·8 \times 10^{-4}$ of the
ground state strength, is from a state at an excitation of 1·8 MeV.

β-DECAY

In Chapter 1, β-decay has been introduced as a
process involving the simultaneous emission of a β-
particle and a neutrino. The neutrino was postulated
to satisfy the requirements of conservation of energy,
momentum and angular momentum in this process. The
β-spectrum is continuous up to an upper limit which is
the energy expected for the β-particle in the absence
of an undetected particle. Similarly experiments involving the direction of recoil of the daughter nucleus indicate that momentum is not conserved - or
preferably that an undetected particle has also been
emitted. Also, to conserve angular momentum a
neutron cannot decay into two spin-$\frac{1}{2}$ particles,
whether isolated or inside a nucleus. The neutrino
must be neutral since charges balance without it. It

must have intrinsic spin $\frac{1}{2}$ and a very small rest
mass, since the β-energy at maximum accounts very
closely for all the energy available. Efforts were
made in the past to detect the neutrino, but failed.
Now that recent efforts have been successful, it
is not surprising that former methods failed since it
is now known that neutrinos can pass through the Sun,
the centre of which is a prolific source - indeed
neutrino emission accounts for a considerable fraction
of the energy emitted by stars.

A simple theory of β-decay was provided by Fermi,
who exploited the similarity with emission of radiation. A proton (or neutron) changes into a neutron
(or proton) by simultaneously creating the β-ν pair
just as an excited state of an atom changes to a
lower state by creating a photon. Note that in isolation a proton cannot decay to a neutron because it
has less mass, but inside a nucleus a bound proton
may be in a state of higher energy than an equivalent
bound neutron, so the process can take place; if the
total masses are favourable then the reaction can go.

Considering first the decay of an isolated
neutron ($n \rightarrow p + e^- + \bar{\nu}_e$) then the wavefunctions of the two
light particles can be represented by plane waves in
the directions of emission, $\psi_\beta = N_\beta \exp(i \, \underline{k}_\beta \cdot \underline{r}_\beta)$ and
$\psi_\nu = N_\nu \exp(i \underline{k}_\nu \cdot \underline{r}_\nu)$. The origin of the coordinate
system is assumed to be the location of the nucleon
both before and after the event (i.e. recoil has been
neglected for simplicity). The plane-wave approximation should be adequate for the neutrino, but
for the β-particle the long-range Coulomb interaction
with the proton (or nucleus when at a later stage the
nucleons are considered to be part of a nucleus) will

distort the wave - this point will be taken up later. In the absence of guidance, Fermi assumed the simplest possible interaction namely that it is proportional to the simultaneous overlap of all four particles, i.e.

$$H_{fi} = g \int \psi_f^* \psi_i \, d\tau = g \psi_\beta^*(0) \psi_\nu^*(0)$$

in this simple case of point nucleons at the origin. From perturbation theory (Fermi's golden rule) the transition probability per unit time is

$$P = \frac{2\pi}{\hbar} |H_{fi}|^2 \frac{dn}{dE} \, ,$$

where dn/dE is the density of final states. In applying this formula it is convenient to enumerate the final states when the system is confined to a specified (large) volume Ω. The normalization factors for the plane waves, N_β and N_ν, are then both $1/\Omega^{\frac{1}{2}}$, and the density of plane-wave states of a particle having momentum between p and $(p+dp)$, with the particle anywhere in Ω is $p^2 dp \Omega / 2\pi^2 \hbar^3$. The number of states for the β-momentum in the range p_β to $(p_\beta + dp_\beta)$ and the ν-momentum in p_ν to $(p_\nu + dp_\nu)$ is then

$$dn = (p_\beta^2 dp_\beta / 2\pi^2 \hbar^3) \cdot (p_\nu^2 dp_\nu / 2\pi^2 \hbar^3) \Omega^2.$$

In multiplying these two functions together it has been assumed that there is no correlation between the directions of emission of electron and neutrino. This is partly because of the initial simple assumptions, but chiefly because the final nucleon is available to take up the requisite momentum required for conservation; because it is so massive it can do this and still have little effect on the energy balance. It does not appear directly in the calculations, but its presence has been felt nevertheless.

To determine dn/dE it is necessary to transform from variables p_β, p_ν to p_β, E, using the equation

$$E = c(p_\nu^2 + m_\nu^2 c^2)^{1/2} + E_\beta,$$

where E is the total available energy. The transformation is

$$dp_\beta dp_\nu = \frac{\partial p_\nu}{\partial E} dp_\beta dE = \frac{1}{v_\nu} dp_\beta dE = \frac{1}{c} dp_\beta dE,$$

since the rest mass of the neutrino is either zero or close to it. Very close to cut-off of the β-spectrum this neglect of the rest mass of the neutrino could distort the spectrum, but such effects have not been observed. Thus the density of states is given by

$$\frac{dn}{dE} = \frac{\Omega^2 p_\beta^2 p_\nu^2}{4\pi^4 \hbar^6 c} dp_\beta = \frac{\Omega^2 p_\beta^2 (E-E_\beta)^2}{4\pi^4 \hbar^6 c^3} dp_\beta,$$

using $E_\nu = cp_\nu = E-E_\beta$.

The transition probability per unit time to produce electrons in the momentum range p_β to $(p_\beta + dp_\beta)$ is

$$P(p_\beta) dp_\beta = \left(\frac{g^2}{2\pi^3 \hbar^7 c^3}\right)(E-E_\beta)^2 p_\beta^2 dp_\beta, \qquad (4.3)$$

or in terms of energy

$$P(E_\beta) dE_\beta = P(p_\beta) \frac{dp_\beta}{dE_\beta} dE_\beta = \left(\frac{g^2}{2\pi^3 \hbar^7 c^3}\right)(E-E_\beta)^2 \frac{p_\beta^2}{v_\beta} dE_\beta.$$
$$(4.4)$$

Note that the result is independent of Ω, the volume
in which the system was confined. In the non-relativistic approximation the energy spectrum is proportional
to $(E - E_\beta)^2 \sqrt{E_\beta}$, and this accounts for the general
shape of a β-spectrum.

To extend this treatment to complex nuclei it·
is necessary to consider the decay of each nucleon
as an independent event and to sum the amplitudes of
each event. So far the interaction is at a point, so
the β-ν combination cannot take away orbital angular
momentum; since no mention has been made of spin the
combination must be assumed to be the spin=0 state.
In the nucleus the decaying nucleon merely changes
its charge but remains in the same nuclear state -
the instantaneous effect can be represented by
$\sum_k \tau_k^\pm \psi_i$, where $\tau_k^\pm$ is the isospin raising or lowering
operator acting on the kth nucleon in the nuclear
wavefunction ψ_i (τ^- for β^- emission, τ^+ for β^+).
This instantaneous wavefunction must be projected
onto the final wavefunction to determine the nuclear
contribution to the process, giving the matrix element

$$M_{fi} = \int \psi^* \sum_k \tau_k^\pm \psi_i \, d\tau.$$

The summation may be taken over all nucleons since τ^-
acting on a proton gives zero - it changes neutrons to
protons - and correspondingly for τ^+. Obviously, from
what has been said above, the selection rule $\Delta J=0$, with
no parity change, holds for the nuclear states.

Extending the theory to include the possibility
of the β-ν combination taking away one unit of spin
leads to a parallel decay obeying the selection rule
$\Delta J=0, \pm 1$, with no parity change (but $J=0 \to 0$ excluded).

The nuclear matrix element must contain a spin operator capable of changing the spin orientation of the
nucleon k - it must be in the form of a vector in order
to take away (spin) angular momentum. There is no need
to go into its form here; the simple picture of the
β-ν combination being emitted in a wave having $L=0$,
$S=1$ is sufficient to arrive at the above selection
rule. The spinless emission is known as the Fermi process and the spin-1 emission is ascribed to Gamow-
Teller. The processes so far discussed are termed
'allowed'.

The treatment of nucleons as point particles is an
approximation which allowed the β-ν combination to
take away no orbital angular momentum. Whilst there
is little to be gained by giving the nucleon structure
in the decay of an isolated neutron (to preserve
parity a two-unit change would be necessary and this
is unlikely for such a small structure), it should be
noted that even for point nucleons it may be possible
to change the orbital angular momentum of a nucleus.
If the emitting nucleon is at the periphery of the
nucleus then, classically, its recoil momentum will
produce a change of angular momentum. The greatest
change will occur when an electron of maximum energy
is emitted tangentially from the surface, and will be
of order $(E_\beta/c)R_N$. Taking $E_\beta \sim 1$ MeV (for which
$E_{\beta/c} = p_\beta$ is only approximate) and $R_N \sim 6$ fm (a medium-
size nucleus) gives $0\cdot03\hbar$ for the classical angular
momentum in this extreme situation. Thus the taking
away of one unit of angular momentum must be looked
upon as possible but improbable. Mathematically, the
two plane waves of β and ν are amalgamated to a single
plane wave $\exp(i\underline{k}.\underline{r})$, where $\underline{k} = \underline{k}_\beta + \underline{k}_\nu$, and, since

the interaction is still assumed to be at a point,
$\underline{r}_\beta = \underline{r}_\nu = \underline{r}$. This plane wave is expanded within the
confines of the nucleus giving
$\exp(i\underline{k}.\underline{r}) = 1 + i\underline{k}.\underline{r} - \frac{1}{2}(\underline{k}.\underline{r})^2$ . . . . The matrix
element for the process will now become

$$H_{fi} = g\int\psi_f^* \sum_k \tau_k^\pm (1 + i\underline{k}.\underline{r}_k \ldots)\psi_i \, d\tau,$$

where for convenience the possible spin factor is
omitted. If ψ_f and ψ_i have opposite parities then
the leading term in the bracket will integrate to
zero, but the next term, which is itself odd, may not
do so. If polar axes are chosen such that $\underline{k}$ lies
along z then $\underline{k}.\underline{r}_k$ becomes $kr_k\cos\theta$. The angular wavefunctions of nucleon k in ψ_f and ψ_i are in the form
of spherical harmonics $Y_{l,m}(\theta,\phi)$, and their properties
are such that $\int Y_{l_f,m_f}^* (\theta,\phi)\cos\theta \; Y_{l_i m_i}(\theta,\phi)d\Omega$ will be
zero unless $m_f = m_i$[†] and $l_f = l_i \pm 1$. Thus the kth
nucleon changes angular momentum by one unit (cf.
emission of a photon by an electron in an atom) and,
by recoupling to the rest of the nucleus, the selection rule follows: $\Delta J=0, \pm 1$, with change of parity
($J=0 \to 0$ excluded) for the Fermi process. By coupling
up to $S=1$, the corresponding Gamow-Teller selection
rule is $\Delta J=0, \pm 1, \pm 2$, with change of parity.

Consideration of the radial integral indicates
that these transitions will be down by a factor $(kR_N)^2$
on the allowed transition - the k^2 term arises from
the integral and the R_N^2 from the fact that an integral
$\int\psi_f^* r\psi_i r^2 dr$ must be $\sim R_N \int |\psi_f| |\psi_i| r^2 dr$, and we assume
that all integrals of the latter type have approximate-
ly the same value. These types of transition are
known as 'first forbidden' since they are down by the

[†]A different choice of axes results in $m_f = m_i \pm 1$.

factor $(kR_N)^2$, which, as previously estimated, is
$\sim 10^{-3}$ or 10^{-4}. The next term in the series would
produce second forbidden transitions and will be down
by the same factor again.

     Before examining whether this classification con-
forms with experimental fact, it is necessary to consider the effect of nuclear charge on the process
since this will vary from nucleus to nucleus. The
wavefunction of a particle in a Coulomb field is complicated, but here we merely need to know the behaviour
at $r = 0$, where the process occurs, and $r = \infty$, where
the products are observed. If at infinity the wavefunction for the β is the original plane wave then
$|\psi_\beta(0)|^2$ will be multiplied by the factor
$F(Z,E) = 2\pi\eta\{1-\exp(-2\pi\eta)\}^{-1}$, where $\eta = \pm\, e^2/4\pi\varepsilon_0\hbar v_\beta$
(+ for electrons, - for positrons), v_β being the speed
at infinity and Z the atomic number of the final nucleus. The Coulomb correction enhances the probability
of electron-emission and decreases that of positron
emission, especially at low energies. In the latter
case it is effectively the Gamow barrier-penetrability
factor since the positron may well be created with
negative kinetic energy and need to penetrate the
barrier before emerging with positive energy. The
enhancement of the β^--process occurs because the
attractive potential increases $|\psi_\beta(0)|^2$ relative to
$|\psi_\beta(\infty)|^2$. The normalization of the wave is largely
determined by $|\psi_\beta(0)|^2$ since Ω is very much larger than
the nuclear volume, but the transition probability
depends on $|\psi_\beta(0)|^2$. The effect of the Coulomb
potential on $\beta^\pm$-spectra is shown in Fig.4.2.

     With the Coulomb correction the momentum spectrum

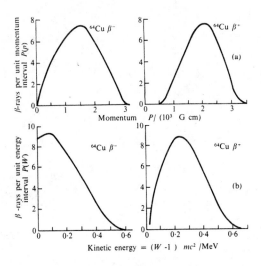

Fig.4.2. Effect of Coulomb potential on β-spectra for ^{64}Cu decay, which shows β$^-$ and β$^+$ (and electron-capture) decay. Note that the β$^-$ energy spectrum has finite intensity at zero kinetic energy, but the momentum spectrum does not (the effect of the relationship $dE = vdp$). (Based on EVANS (1955). *The atomic nucleus*, McGraw-Hill, New York.)

for allowed transitions is of the form

$$P(p_\beta)\,dp_\beta = CF(Z,E_\beta)(E-E_\beta)^2 p_\beta^2 dp_\beta, \qquad (4.5)$$

where the constant C contains terms independent of the β-energy. Thus a plot of $\{P(p_\beta)/Fp_\beta^2\}^{1/2}$ against E_β is a straight line intersecting the abscissa at E. This, or its equivalent for the energy spectrum, is known as a Kurie plot, an example of which is shown

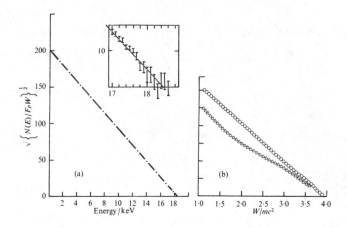

Fig.4.3. (a) Kurie plot for allowed decay of tritium. Inset
         shows fit near end point. Lack of deviation indicates
         that the rest mass of the neutrino must be small.
         (Based on LEWIS (1970). *Nucl.Phys.* A151, 120.)
         (b) Kurie plot for first forbidden decay of ^{89}Sr. The
         curved plot is the usual plot with F_0 in the ordinate
         function; the linear plot contains a theoretical cor-
         rection for the first forbidden transition. (Based on
         WOHN and TALBERT (1970). *Nucl.Phys.* A146, 33.)

in Fig.4.3. The linear plot results for allowed transitions only; in forbidden transitions terms in
$k^2 = k_\beta^2 + k_\nu^2 + 2\underline{k}_\beta.\underline{k}_\nu$ appear, which depend upon the
relative direction of the β-ν. On integrating over
all directions, the term $\underline{k}_\beta.\underline{k}_\nu$ averages to zero, and
the two remaining terms inject a factor $\{p_\beta^2 + (E-E_\beta)^2/c^2\}$
into the momentum spectrum for a first forbidden
transition. The Kurie plot will not be linear, but
a modified Kurie plot, including this factor, can

restore the straight line. This technique can be used
to determine the type of decay (see Fig.4.3).

A convenient classification of β-transitions is
based on the lifetime of the decay

$$\frac{1}{\tau} = \int P(p_\beta) \, dp_\beta = g^2 |M_{fi}|^2 f,$$

which defines f as an integration over all the energy
dependent terms in the spectrum and contains all the
fundamental constants excluding the coupling constant.
Thus for an allowed transition

$$f_0 = (2\pi^3 c^3 \hbar^7)^{-1} \times \int_0^{p_{max}} (E - E_\beta)^2 p_\beta^2 F(Z, E_\beta) \, dp_\beta.$$

The product $f_0 t_{\frac{1}{2}}$, where $t_{\frac{1}{2}}$ is the half-life ($= \tau \ln 2$),
is therefore a measure of $(g^2 |M_{fi}|^2)^{-1}$ for an allowed
transition and is found to vary between 10^3 and 10^6
in magnitude. Transitions having values near 10^3 are
known as 'super-allowed' since they correspond to
(almost) complete overlap of the nuclear wavefunctions.
They are usually Fermi transitions between states of
the same isospin multiplet (e.g. β-decay between mirror
nuclei), but there exist a few Gamow-Teller transitions in very light nuclei where the overlap is also
complete. The remainder of the range corresponds
to a diminished overlap. The *same* product $f_0 t$ for
a first forbidden transition will be of the form
$\{g^2 (|M_{if}|^2 / R_N^2) \overline{(kR_N)^2}\}^{-1}$, where the bar denotes a suitable average over the β-spectrum. For complete over-
lap this is up by a factor $(kR_N)^{-2}$ in value on an
allowed transition giving a range of order $10^6 - 10^{10}$.
Similarly, a second forbidden transition will give
$f_0 t$ values $> 10^{10}$. Thus there is no strict division

between the different classes, but for certain ranges
of $f_0 t$ values it is possible to deduce the type with
a reasonable degree of confidence.

Note that these last considerations should really
apply only to β^- transitions since, as has been mentioned in Chapter 1, an alternative decay mode to β^+
is available. This will be discussed briefly in the
next section.

ELECTRON CAPTURE

Since atomic nuclei are in general surrounded by
atomic electrons an alternative process to β^+-decay
can take place, namely,

$$p + \beta^- \rightarrow n + \nu_e \qquad \text{(within the nucleus)}.$$

This process is of greatest importance at low available energies - indeed there is a range of available
energy (see Chapter 1) of $0\text{-}1\cdot02$ MeV over which β^+-
decay is impossible and e^--capture only can take place.
Also it is expected to be largest for K-electrons (provided the energy balance is not critical), since they
have strongest overlap with the nucleus, and for nuclei of large Z, since the electron orbits are more
tightly bound, again leading to greater overlap.

Applying Fermi's golden rule to this process,
the density-of-states function will be that of the
neutrino alone - the nucleus is constrained to recoil
in the opposite direction and both particles have
fixed energy. The probability of finding the electron
at the origin of the coordinate system is now no longer
inversely proportional to the volume of the enclosure
but to the volume of the K-orbit (for K-capture), so

once again Ω cancels (it has also disappeared from the density-of-states function for the electron which is now unity - a defined state). If E is used, as before, to denote the maximum available energy for the β^+-decay, then the energy available to the neutrino will be $(E + 1\cdot02 - B_K)$ MeV, where B_K is the binding energy of the K-electron. There is now no integral to perform since no variation is allowed of the products, and

$$\frac{1}{\tau_K} = \frac{2\pi}{\hbar} \, g^2 |M_{if}|^2 \, \frac{1}{\pi} \left(\frac{Z}{a}\right)^3 \frac{(E+1\cdot02-B_K)^2}{2\pi^2\hbar^3 c^3} \, , \qquad (4.6)$$

where a is the (hydrogen) Bohr radius, so that a/Z is the mean radius of the K-electron wavefunction in the parent nucleus, and where the matrix element M_{if} is the same as before, and $1/\tau_K$ is the probability per second of K-capture. Thus the probability of the process increases as Z^3, whilst that for β^+-decay decreases with Z (due to the Coulomb effect); for light nuclei β^+-decay will dominate except for an energy range in E from the negative value $(-1\cdot02 + B_K)$ MeV to slightly above zero. But for high Z, K-capture is overwhelmingly predominant for β^+-energies commonly encountered.

RELATIVISTIC THEORY AND PARITY VIOLATION

The account of β-decay so far described has been essentially non-relativistic (NR) in its approach, apart from the use of a relativistic equation connecting momentum and energy - a trivial aspect. For relativistic invariance it is essential that time and space be treated on an equal footing, and this is not the case with the Schrödinger wave equation, which is

of second order in spatial derivatives and first
order in the time derivative. Putting $E = i\hbar \; \partial/\partial t$,
$p_x = (\hbar/i) \; \partial/\partial x$ into the relativistic (R) equation
$E^2 = c^2 p^2 + m_0^2 c^4$ gives the Klein-Gordon equation which
is second order in all four derivatives. However,
its solutions are found to conform to Bose statistics,
e.g. photons, and the equation cannot be used for
spin-$\frac{1}{2}$ particles. Dirac looked into the conditions
under which this second-order equation factors out
into two first-order equations and further examined
the solutions of the resulting equation. He found
that the solutions were four-component vectors in a
space which was closely allied to the two component
space of a NR spin-$\frac{1}{2}$ particle. Indeed two of the
components are down by a factor v/c on the other two
such that in the limit $v \to 0$, the R system reduces to
the NR system. In addition the 'small' components
have opposite parity to the 'large' components, creating the possibility of parity violation at relativistic energies.

In β-decay, the approximation $v/c \ll 1$ is rarely
valid for the electron and never so for the neutrino
- even for tritium decay the β-particle has maximum
kinetic energy of 18·6 keV corresponding to $v/c \sim 0·27$.
Thus the product of $\psi_\beta \psi_\nu$, which in the NR limit gives
four terms which can be regrouped into a scalar ($S=0$)
and a three-vector ($S=1$) gives in the R limit six-
teen terms which can be regrouped into two scalars,
two four-vectors, and a six-component antisymmetric
tensor. The two vectors are respectively polar and
axial; the former changes sign under the parity opera-
tion whilst the latter does not, e.g. $\underline{r}$ and $\underline{p}$ are
polar three-vectors whilst $\underline{r} \wedge \underline{p}$ is an axial

three-vector. Of the two scalars one is a pseudo-
scalar in that it changes sign under the parity operation; an example of such a scalar is the triple three-
vector product $\underline{a}.(\underline{b}\wedge\underline{c})$, where $\underline{a}$, $\underline{b}$, $\underline{c}$ are polar three-
vectors. Such a product can be used classically
(e.g. to represent a volume) but the sign property
is usually ignored.

Thus a full relativistic treatment is much more
complicated than the picture presented here and has
built into it terms which mix the parity of the relativistic components. It is possible to choose an
interaction in such a way as to eliminate the parity-
violating component, and this was believed to be the
case until specific measurements were made to test
parity conservation in β-decay. Such measurements
are designed to determine whether pseudo-scalars are
necessary in the description of the reaction. Such
a scalar product is $\underline{\sigma}.\underline{p}$, where $\underline{\sigma}$ represents the NR
spin of the electron and $\underline{p}$ its momentum or $\underline{J}.\underline{p}$, where
$\underline{J}$ is the spin of the initial nucleus and the β-momen-
tum $\underline{p}$ is now measured relative to the same axis as $\underline{J}$.
Obviously, if parity is conserved then $\langle\underline{\sigma}.\underline{p}\rangle$ and
$\langle\underline{J}.\underline{p}\rangle$ must be zero. The first measurements were made
on the decay of ^{60}Co, with the initial state polarized
by cooling in a magnetic field. It was then found
that the number of β-particles directed into the hemisphere with a component along the field was different
from that directed into the opposite hemisphere, in-
dicating that $\langle\underline{J}.\underline{p}\rangle$ was not zero. The final conclusion
to be drawn from such measurements was surprising,
not so much because parity was not conserved but because its violation appeared to be the maximum possible when extrapolated to the limit $v \rightarrow c$.

When polarization measurements are ignored as in the simple determination of the spectrum, then the simple predictions of the Fermi theory are largely correct, but, in more detail, separation into 'large' and 'small' terms of the nucleon wave functions is necessary. Part of a first forbidden transition can arise from the 'small' term of an allowed transition, and vice versa, though in the latter case the effect will not be discernible since it is also reduced by the factor $(kR_N)^2$ relative to the main component; in the former case the factor serves to enhance this extra component. Thus, whilst under most circumstances the simple theory copes with allowed transitions, great caution is needed in extending its results to forbidden transitions.

PROBLEMS

4.1. Solve the one-dimensional problem of transmission through the rectangular potential as given in Fig.4.1 (p.91). Show that in the limit of the barrier width approaching zero, the transmission becomes that calculated for a potential step.

4.2. Solve the integral $\int_{r_n}^{r_0} \gamma \mathrm{d}r$ (see p.95) for the more realistic potential of Fig.4.1., where the limits of integration are the classical distance of closest approach and the nuclear radius (use the transformation $r = r_0 \cos^2\theta$). Show that, in the limit $E_\alpha \ll$ barrier height, the Gamow factor $\exp(-2\pi Zze^2/4\pi\varepsilon_0\hbar v)$ is the important term in s-wave barrier penetration.

4.3. Show that the non-relativistic formula for the β-spectrum for an allowed transition gives a

mean kinetic energy of the β of $\frac{1}{3}$ the maximum
energy, whilst the relativistic formula, for
a very high maximum energy, gives a factor $\frac{1}{2}$.
Derive expressions for the maximum recoil energy
of the nucleus of mass A when the β is emitted
at the mean energy in both cases. (Neglect Coulomb effects.)

4.4*. Derive an expression for the electron momentum
spectrum in allowed β-decay (ignore Coulomb cor-
rections). Find the approximate dependence of
the total decay rate on the maximum electron
momentum p_0 when $p_0 \gg mc$.

   Compute the value of the partial decay rate
for the pion decay $\pi^+ \rightarrow \pi^0 + e^+ + \nu_e + 4 \cdot 5$ MeV
given that the Fermi decay of ^{14}O to ^{14}N (excited)
has a momentum end point $p_0 = 2 \cdot 26$ MeV/c and a
mean lifetime of 103 s.

4.5*. Explain the meanings of the following terms: (a)
first forbidden, allowed and super-allowed β-
transitions, and (b) Fermi and Gamow-Teller matrix
elements. Classify with regard to (a) and (b) the
following: (i) $n \rightarrow p$; (ii) $^6_2He(0^+) \rightarrow ^6_3Li(1^+)$;
(iii) $^{14}_8O(0^+) \rightarrow ^{14}_7N(0^+)$, $f_0t = 3 \cdot 3 \times 10^3$;
(iv) $^{35}_{16}S(\frac{3}{2}^+) \rightarrow ^{35}_{17}Cl(\frac{3}{2}^+)$, $f_0t = 1 \cdot 0 \times 10^5$;
(v) $^{36}_{17}Cl(2^-) \rightarrow ^{36}_{18}Ar(0^+)$.

4.6. ^7_4Be decays by electron capture chiefly to the
ground state of ^7_3Li, and the recoil energy of
the ^7Li has been measured to be 57 eV to 1 per
cent accuracy. If the mass difference between
the two atoms is $0 \cdot 862$ MeV/c^2 show that the re-

sult is consistent with a neutrino rest mass anywhere in the range of zero to $\sim$ 10 keV/c^2.

5. Spontaneous decay of nuclei II: electromagnetic transitions

Thus far the discussion has been restricted to the decay of nuclear ground states, apart from the occasional intrusion of an excited state decaying, for example, as an isometric state, or giving a 'long-range' α-particle. In the next chapter states at high enough excitation to emit nucleons will be considered, but here the electromagnetic decays of states within 1 MeV or so above the ground state are treated. The electromagnetic process is, in general, a stronger one than the weak β-process, so that such excited states only rarely β-decay, though there are a few instances when the excited states lie close to the ground.

The treatment of radiation presented here is largely based upon the corresponding process of electric dipole radiation from excited atomic states. Later in the chapter it will be shown that such radiation can take place only between states satisfying the criteria $\Delta J = 0, \pm 1$ and change of parity. But, experimentally, using techniques to be described briefly in the next chapter, it is found that in nuclei γ-transitions frequently take place between states not conforming to these criteria. It is therefore necessary to extend the treatment to cover the case of emission via the higher electric multipoles and via magnetic multipoles. Such transitions can be ignored in atoms since they occur so seldomly and then only under very special conditions. If the rapid process of electric dipole (E1) radiation cannot take place,

then the state usually de-excites by collision.   In
nuclei, on the other hand, it transpires that E1
emission is not a great deal faster than the other
processes, and in any case most low-lying states
have the same parity as each other since they probably arise from different couplings of the same
shell model particles or from collective motion.
But nuclei are isolated from each other by virtue of their charge, and are coupled only weakly

to the atomic electrons; a state once formed can only
decay via the electromagnetic process, though it need
not necessarily lead to γ-emission since the Coulomb
coupling to the atomic electrons can result in the
emission of an electron termed, misleadingly, a
'conversion electron'. This latter process is electromagnetic and gives the same information concerning
nuclear states as does γ-emission.

γ-RADIATION

The difficulty in calculating the spontaneous
decay of an excited state is that one does not know
the perturbing term in the wave equation which is
responsible for the transition. However, the reverse
process of absorption can be treated by using the
electric vector of the radiation as a time-dependent
perturbation. Einstein, using a statistical argument
(known as 'Einstein's As and Bs'), obtained a relationship between the two processes. A fuller account is given in Pauling and Wilson (1935), but
very briefly the argument is as follows.

The coefficient of absorption is defined such
that the probability of a transition from a lower
state n to a higher state m is given by $B_{n \to m} \, \rho(\nu_{mn})$,

where $\rho(\nu_{mn})$ dν is the energy density of radiation in the range ν_{mn} to ν_{mn} + dν) with ν_{mn} the frequency of the transition, i.e. $\nu_{mn} = (E_m - E_n)/h$. The probability of transition down is given by $A_{m \to n} + B_{m \to n}\rho(\nu_{mn})$, where $A_{m \to n}$ is the coefficient of spontaneous emission and $B_{m \to n}$ is the coefficient of induced emission. If we consider the equilibrium of this two-state system in a bath of radiation at infinite temperature then $B_{n \to m} = B_{m \to n}$, since the two states will be present in equal proportions, and the induced processes will be much more probable than the spontaneous process. It is assumed that the two states are non-degenerate; this can be achieved by applying a magnetic field to separate out sub-states of n and m - the transitions between a single sub-state in each level is then under consideration. If now the equilibrium is considered at a finite temperature, then the ratio of abundances will be given by the Boltzmann factor, $N_m/N_n = \exp\{-(E_m - E_n)/kT\}$, whilst $\rho(\nu)$ is given by Planck's radiation law,

$$\rho(\nu) = (8\pi h \nu^3/c^3)\{\exp(h\nu/kT) - 1\}^{-1} .$$

But at equilibrium

$$N_n/N_m = \{A_{m \to n} + B_{m \to n} \rho(\nu_{mn})\}\{B_{n \to m} \rho(\nu_{mn})\}^{-1} .$$

From these equations

$$A_{m \to n} = (8\pi h \nu_{mn}^3 /c^3)B_{n \to m}$$

$$= (2\hbar \omega_{mn}^3/\pi c^3)B_{n \to m}$$

It is not proposed to go through the complete calculation of $B_{n \rightarrow m}$ as given in the above reference for El transitions, but to show how to modify this approach in order to produce the other multipoles. The perturbation resulting in absorption was taken to be due to the electric vector of the radiation field, $\underline{E}_0 \exp\{i(\underline{k}.\underline{r}-\omega t)\}$. Since $k = mv/\hbar = E/\hbar c$, for a photon of ~ 1 MeV and a nuclear radius of 5 fm, then $kR_N \sim \frac{1}{40}$. Thus the simplification $E_0 \exp(-i\omega t)$ is a fairly good one, though not so good as in the atomic case. Now a uniform electric field is given by a dipole at infinity, so this approximation, giving a perturbing potential $- e\underline{E}_0.\underline{r} \exp(-i\omega t)$, leads to El absorption of amplitude proportional to the matrix element $\int \psi_m^* (\sum_i ez_i) \psi_n d\tau$, where the summation i is taken over the protons of the nucleus and where E_0 has been taken as directed along the z-axis. This gives

$$A_{m \rightarrow n} = \left(\frac{4\omega_{mn}^3}{3\hbar c^3} \right) \left| D_{mn}^{(z)} \right|^2 = \left(\frac{4\omega_{mn}^3}{9\hbar c^3} \right) \left| D_{mn} \right|^2, \qquad (5.1)$$

where in the latter expression averaging over all directions of E_0 has been made in forming the dipole matrix element - and also averaging over all substates m and n, which is correct for the initial state but not for the final state. A further factor $(2J_n + 1)$ should therefore be present, but it is a degree of refinement which will be ignored. The suitably averaged dipole matrix element becomes e times the radial integral contribution to the moment for a particular choice of sub-states.

The higher electric multipole contributions can be obtained by using the expansion $\exp(i\underline{k}.\underline{r}) = 1 + i\underline{k}.\underline{r} - \frac{1}{2}(\underline{k}.\underline{r})^2$ ..... in the expression for the radiation field. The leading term has given a component of the E1 matrix element; the next term gives a component of the E2 matrix element $\int \psi_m^* k e (\sum_i x_i z_i) \psi_n d\tau$ if the propagation vector $\underline{k}$ is taken along the x-axis (it must be perpendicular to $\underline{E}_0$, since electromagnetic waves are transverse). And similarly for higher orders.

From the matrix elements the selection rules arise. Since z is an odd function, the product $\psi_m^* \psi_n$ must also be odd to give a non-zero dipole term, i.e. there must be a change of parity. Similarly, xz is even on reflection through the origin, so the E2 transition is characterized by no change of parity. In more detail, if the particle making the transition is in a state of orbital angular momentum $(l_i m_i)$ before and $(l_f m_f)$ after the transition, then the angular dependence of the above component of the E1 matrix element will be $\int Y_{l_f m_f}^* (\theta, \phi) \cos \theta \ Y_{l_i m_i} (\theta, \phi) d\Omega$, and the properties of the spherical harmonics (Pauling and Wilson 1935) are such that this integral is zero unless $m_f = m_i$ and $l_f - l_i = \pm 1$ (there are two other components which lead to $m_f - m_i = \pm 1$). Similarly the above component of the E2 matrix element, $\int Y_{l_f m_f}^* (\theta, \phi) \cos \theta \sin \theta \cos \phi \ Y_{l_i m_i} (\theta, \phi) d\Omega$, will be zero unless $m_f - m_i = \pm 1$ and $l_f - l_i = 0, \pm 2$, with $0 \to 0$ excluded since it cannot satisfy the former condition. Again, there are other components, for which $m_f - m_i = 0, \pm 2$, e.g. the xy component giving $\Delta m = \pm 2$.

For nuclei consisting of a single proton outside closed shells, the selection rules on (J, M) will be the same as those obtained on (l, m) for transitions between states corresponding to excitations of this proton, apart from the addition of a spin vector. In the general case it is necessary to decouple the rest of the nucleus from the particle making the transition and then to recouple after the transition. The mathematics of these coupling operations is beyond the scope of this book, but it is not difficult to see that the selection rules will become

E1 : $\Delta J = 0, \pm 1$ ($0 \to 0$ excluded), with parity change,

E2 : $\Delta J = 0, \pm 1, \pm 2$ ($0 \to 0, \frac{1}{2} \to \frac{1}{2}$ excluded), with no parity change.

The exclusions arise because all components of a given moment must be allowed since they transform into each other on rotation of axes. Thus $\Delta m = 2$ (and therefore $\Delta M = 2$) must be possible in E2, and it is patently not so for the excluded transitions.

So far no mention has been made of magnetic multipole transitions for the very good reason that the interaction has been taken to be that of an electric field. But electromagnetic radiation also possesses a magnetic vector $\underline{H}_0$, perpendicular to both $\underline{E}_0$ and $\underline{k}$. Again the leading term will arise by assuming the (magnetic) field constant over the nucleus giving rise to the perturbation term $-\underline{\mu} \cdot \underline{B}_0 \exp(-i\omega t)$, where the magnetic moment $\underline{\mu} = \beta_N \sum_i (g_s^i \underline{s}_i + g_l^i \underline{l}_i)$, where β_N is the nuclear magneton and g_s^i, g_l^i are the appropriate spin, orbital g-factors for the nucleon i. To simplify matters, consider the case of a single neutron outside a closed-shell core then
$\underline{\mu} \cdot \underline{B}_0 = \hbar^{-1} g_s \beta_N (s_x B_{0x} + s_y B_{0y} + s_z B_{0z})$, where s_x, s_y, s_z are spin operators (see Appendix D). If the nucleus

conforms to the J-J coupling scheme (and most do, at
least approximately, and especially near closed shells)
then the two states $J = l \pm \frac{1}{2}$ are distinct and ortho-
gonal. They can be expressed in terms of orbital and
spin product wavefunctions thus:

$$|l+\tfrac{1}{2},m\rangle = \alpha|l,m+\tfrac{1}{2}\rangle|\tfrac{1}{2},-\tfrac{1}{2}\rangle + \beta|l,m-\tfrac{1}{2}\rangle|\tfrac{1}{2},+\tfrac{1}{2}\rangle,$$

$$|l-\tfrac{1}{2},m\rangle = -\beta|l,m+\tfrac{1}{2}\rangle|\tfrac{1}{2},-\tfrac{1}{2}\rangle + \alpha|l,m-\tfrac{1}{2}\rangle|\tfrac{1}{2},+\tfrac{1}{2}\rangle,$$

where α and β are normalized such that $|\alpha|^2+|\beta|^2 = 1$
(they are known, but the values are not needed here).
Since $s_z = \frac{1}{2}\hbar\left(\begin{smallmatrix}1 & 0\\0 & -1\end{smallmatrix}\right)$ it follows that $s_z|l+\tfrac{1}{2},m\rangle$ contains
both $|l+\tfrac{1}{2},m\rangle$ and $|l-\tfrac{1}{2},m\rangle$ as also does $s_z|l-\tfrac{1}{2},m\rangle$. In
treating s_x and s_y it is simpler to recombine them
to form $s^+ = s_x+is_y$ and $s^- = s_x-is_y$ with weighting
factors which will depend upon B_{0x} and B_{0y}. Now,
from Appendix D, $s^+|l+\tfrac{1}{2},m\rangle \propto |l,m+\tfrac{1}{2}\rangle|\tfrac{1}{2},+\tfrac{1}{2}\rangle$, which is
a linear combination of $|l+\tfrac{1}{2},m+1\rangle$ and $|l-\tfrac{1}{2},m+1\rangle$; and
similarly for s^-. Thus the effect of $\underline{s}.\underline{B}_0$ is to mix
the original state $|j,m\rangle$ into states belonging to
both couplings of $\underline{l}$ and $\underline{s}$ and also to mix the orientations, $\Delta m = 0, \pm 1$. Thus the matrix element will
be non-zero if the final state contains one or other
of these components. A similar argument applies for
the term in $\underline{l}.\underline{B}_0$ for the proton. The selection rule
for magnetic dipole single-particle transitions is

M1 : $\Delta l = 0$, $\Delta m = 0, \pm 1$, $|\Delta j| = 0, 1$, no change
of parity,

where $|\Delta j| = 1$ allows the alternative coupling scheme.
In the general case it is necessary to decouple and

recouple with the rest of the nucleus giving

M1 : $\Delta J = 0, \pm 1$, no change of parity, (0→0 excluded

The next step would be to expand the exponential again to give higher magnetic multipoles, but we shall leave it at this point, except for observing that the parity will alternate with the higher magnetic multipoles, as for the electric. It is interesting to note that, whilst E1 transitions are essentially from one shell to another, M1 and E2 transitions can be made between states with the same basic shell structure. In fact most low-lying nuclear states are different configurations of the same basic states so one encounters M1 and E2 transitions much more often than E1.

From the foregoing it is possible to make crude estimates of the ratios of these three main types of transition. E2 is seen to be down by a factor $(kR_N)^2 \sim 1/2000$ on E1 for a medium-weight nucleus, whilst

$$(M1/E1)^{1/2} \sim \hbar^{-1}\beta_N g_s B_0 \int \psi_m^* s_z \psi_n \, d\tau / eE_0 \int \psi_f^* z \psi_i \, d\tau$$

$$\sim (\beta_N g_s / 2eR_N)(B_0/E_0)$$

$$= (\hbar/McR_N)(g_s/4) \ .$$

From the uncertainty principle $\hbar/R_N \sim Mv$ leading to a ratio of intensities $\sim v^2/c^2$ which, for a typical kinetic energy inside the nucleus of order 50 MeV, takes the value $\sim \frac{1}{10}$. This is perhaps an overestimate, a factor $\sim \frac{1}{100}$ is closer.

Blatt and Weisskopf (1952) make estimates of the

matrix elements under simple assumptions arriving at $\frac{3}{4}R_N$ for E1 and $\frac{3}{5}R_N^2$ for E2. For M1 they take $M1/E1 \sim 10(\hbar/McR_N)^2$, where the factor 10 allows for the large-spin g-factors of proton and neutron. Inserting these estimates and the values of the physical constants gives

$$\Gamma_\gamma(E1) = 0 \cdot 07 E_\gamma^3 A^{2/3}$$

$$\Gamma_\gamma(E2) = 4 \cdot 9 \times 10^{-8} E_\gamma^5 A^{4/3},$$

$$\Gamma_\gamma(M1) = 0 \cdot 021 E_\gamma^3 , \qquad (5.2)$$

where Γ_γ is in eV, E_γ in MeV, and A is the mass number. The radiation width Γ_γ is connected with the decay half-life by $t_{1/2}(s) \sim 4 \cdot 6 \times 10^{-16}/\Gamma_\gamma$ (see also Fig. 5.1).

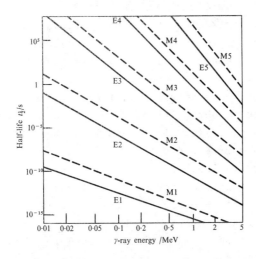

Fig.5.1. Lifetime-energy relationships for γ-radiation, based on Weisskopf's single-particle estimates.

Thus it is seen that the single-particle estimates lead one to expect E1 transitions to be stronger than M1 or E2 at $\sim$ 1 MeV. But, as remarked previously transitions between single-particle states will rarely be found at such low excitation. In addition, there appears to be a tendency for neutrons and protons to move together in nuclei, and, from the remarks below concerning centre-of-mass effects, this will very much reduce the effective charge of the radiating system for E1 transitions. Except for a few transitions in the light elements low-energy E1 transitions are many orders of magnitude weaker than single particle. Since M1 transitions are expected between similar components in complex wavefunctions, they are rarely found to approach single-particle strength, and are typically a few per cent of this value. On the other hand, E2 transitions are often found to be a factor of 100 stronger than the single-particle estimate. The matrix elements so far discussed have been treated implicitly in terms of the shell model in which only a small number ($\sim$ 1) of particles can find suitable initial and final configurations. If, however, all Z protons move coherently, then it is possible in principle to obtain transitions up to Z^2 stronger than single particle. In a rotational band the states have such a correlated motion in which at least 20 nucleons are taking part.

The switch from protons to nucleons is deliberate; it is found that neutron transitions are as strong as proton transitions for both E1 and E2, though for different reasons. In the former case it is because of the finite mass of the nucleus; when a proton moves outward, the rest of the nucleus must move in the opposite direction to keep the centre of mass unchanged. By simply taking first moments (in the classical sense) about the centre of mass it is seen that the system behaves as though it were infinitely massive provided that the charge on the proton is taken as $+e(1 - Z/A) \sim + \frac{1}{2}e$. A similar argument gives the effective charge on the neutron as $-eZ/A \sim -\frac{1}{2}e$. Since first moments were taken this applies to E1 transitions. Performing the same operations with second moments gives only small departures from $+e$ and zero for the two nucleons. Thus the centre-of-mass effect does not produce large E2 neutron transitions. That E2 neutron transitions are as large as proton transitions even in nuclei like ^{17}O and ^{17}F (closed shell plus one neutron or proton) must be ascribed to a collective polarization of the core, and in both cases it must be a large effect.

Since E2 are in general larger than single particle and M1 in general smaller, they are found to be comparable in strength. The commonest transitions in nuclei are therefore found to be M1 or E2 or mixed M1/E2 transitions; in the latter case either type can be larger though with a tendency towards E2 in the heavier, distorted nuclei and to M1 in the lighter nuclei or nuclei near to closed shells.

INTERNAL CONVERSION

It is possible for an excited nucleus in an atom to de-excite by emitting an atomic electron by an electromagnetic process known as internal conversion. It has an atomic counterpart known as the Auger effect when, as an alternative to X-radiation, an outer-shell electron is emitted instead. This process could be

looked upon merely as the photoelectric effect in
which the atom happens to be the same one as contains
the radiating nucleus; indeed, such a process must
occur and is more likely to occur for this particular
atom than any other atom, since the radiation density
is vastly greater than for any other atom. However,
the photoelectric effect is not very probable for 'any
other atom', so even for the particular atom it is
likely to account only for a very small fraction of
decays. But internal conversion can account for the
major fraction of low-energy transitions in heavy
nuclei, so a single-stage process is indicated in
which the energy is transferred directly from the
nucleus to the ejected electron. The Coulomb field
provides the perturbation for the process.

The Coulomb potential at a given point $\underline{r}_e$ in
space due to the nucleus is of the form

$$V(\underline{r}_e) = \sum_i \frac{e}{|\underline{r}_e - \underline{r}_i|} \frac{1}{4\pi\varepsilon_0} ,$$

where the summation is taken over the protons, of
position vector $\underline{r}_i$, inside the nucleus. If the field
point $\underline{r}_e$ lies outside the nucleus, i.e. $|\underline{r}_e| > |\underline{r}_i|$
for any $\underline{r}_i$ inside the nucleus, then the term $1/|\underline{r}_e - \underline{r}_i|$
can be expanded in powers of $\underline{r}_i \cdot \underline{r}_e / r_e^2$. When r_i is
averaged over the nucleus each term in the expansion
produces a nuclear electric multipole moment just as in
the corresponding expansion of the radiation field. There
is also a similar expansion in terms of magnetic multipoles arising from the magnetic interaction between the
nuclear magnetic moment and that of the electron. Thus
γ-emission and internal conversion depend upon the
same nuclear property.   Without going into details,

the process is expected to be more probable for the
heavier nuclei, since the electron being ejected is
then on average, closer to the nucleus and subject to
a larger electromagnetic field. Also from considerations of density of final states, the process is expected to be larger, relative to γ-emission, for lower
transition energies, provided they exceed the binding
energy of the electron. The heavier the particle
the greater its momentum for a given energy (assuming
that energy is not required to create the particle)
and photons have zero rest mass. The ratio of densities of states favours internal conversion most
strongly just above threshold and least strongly at
high energies. The details of the process also favour
conversion just above threshold.

The internal conversion coefficient α is defined
as the ratio of probabilities of the conversion process to γ-emission. It can be further divided into
components $\alpha = \alpha_K + \alpha_L + \alpha_M$, etc. corresponding to
the shell from which the electron is emitted. Its
behaviour is shown graphically in Fig.5.2. The coefficients α_K, α_L, α_M, etc. have been extensively
tabulated for all atoms over a wide range of ener-
gies and multipolarities. Measurements of α, α_K,
α_L, α_M are used to determine the type of transition
and even the amount of mixing in a mixed transi-
tion. Since the process is one of two-body break-
up, the energy spectrum consists of discrete
lines separated by the binding-energy differences
of the atomic shells. Such lines were first discovered when superimposed on continuous β-spec-
tra and initially led to confusion in the interpretation of β-spectra.

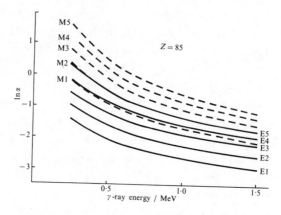

Fig.5.2. Internal conversion coefficients (α) as functions of
energy and multipolarity for a heavy element. (Based
on SEGRE (1959). *Experimental nuclear physics*, VOL.III,
Wiley.)

$J = 0 \rightarrow 0$ TRANSITIONS

It is not strictly correct to say that the conversion process has exactly the same nuclear dependence as that of γ-emission. The statement arose
because of the similarity of two expansions and the
expansion of $1/|\underline{r}_e - \underline{r}_i|$ is of the stated form only if
$\underline{r}_e$ lies outside the nucleus. Whilst an atomic electron is usually well outside the nucleus, it does
overlap it for a small fraction of the time; when
$|\underline{r}_e| < |\underline{r}_i|$ the expansion becomes quite different, and
in particular the monopole component, which for an
external point gives a time-independent field because
centre of charge and centre of mass coincide, can now
have a time variation. Looked at simply, a monopole
oscillation of the nucleus is a pulsating mode, having
no effect on an external point, but for an internal
point the potential depends upon the instantaneous
nuclear radius and this is oscillating in time.

Although the finite size effect is small for transitions capable of radiating, for $0 \to 0$ transitions it is crucial. There are examples of such transitions of less than 1 MeV which can de-excite by no other process - the simultaneous emission of two γ-rays is possible but has never been detected: An example is the first excited state of ^{72}Ge (0^+, 690 keV) which decays in this way with a half-life of 0·3 μs. Above 1 MeV the process in the next section becomes possible.

INTERNAL PAIR CREATION

In this process a state of excitation greater than $2 m_e c2$ (1·02) MeV) can de-excite by the creation and emission of an $e^+ - e^-$ pair. Since the emission can take place from inside the nucleus, $0 \to 0$ transitions can occur this way. Also the density of final states will increase more rapidly with energy for this three-body break-up than for internal conversion, so at high enough energies this process will dominate in $0 \to 0$ transitions. An example is the first excited state of ^{16}O (0^+, 6 MeV) which decays by pair-emission in 5×10^{-11}s. Pair-emission in competition with γ-emission has also been observed. Again in principle it should dominate at high energies, from phase-space arguments, but long before it does so the levels have become virtual with many channels open for the emission of nucleons.

PROBLEMS

5.1. Estimate the lifetime of the first excited state of a (spinless) proton in a cubic box of side 5 fm. (Consult Pauling and Wilson (1935) for the quantum-mechanical problem.)

5.2*. What is meant by multipolarity of electromagnetic radiation? Explain why multipolarities other than E1 are commonly observed in nuclei but only rarely in atoms.

The decay scheme of low-lying levels in ^{14}N is (A; 0; 1^+, 0; -), (B; 2·3; 0^+, 1; A, 100 per cent), (E; 5·1; 2^-, 0; A, 67 per cent; B, 33 per cent), (C; 3·9; 1^+, 0; B, 96 per cent; A, 4 per cent) (D; 4·9; 0^-, 0; A 100 per cent), (E; 5·1); 2^-, 0; A, 67 per cent; B, 33 per cent), where the capital letter designates a level, of energy in MeV given next, followed by J^π, T, and finally the γ-decay of the level in percentages to the levels below. Sketch the level scheme and the transitions. Explain the significance of J^π, T and give the possible and expected multipolarities of transitions. Comment on the absence of the transitions DC, DB, ED, and EC. What other decay processes can occur?

5.3. $^{160}_{65}$Tb decays by β-emission to states of $^{160}_{66}$Dy. Two of the γ-rays resulting are associated with an internal-conversion line spectrum as follows: 32·9, 78·0, 84·7, 86·1, 143·0, 188·6, 195·1, 196·5 keV. The K,L.M,N edges in Dy are 53·8, 8·6, 1·9, and 0·4 keV respectively. Find the γ-energies, identifying each conversion line with the nuclear transition and the atomic shell from which it arises. (Ignore the splitting of the L,M,N levels.)

5.4. Determine the effective charges due to centre-of-mass effects for neutron and proton E1 and E2

transitions in $N = Z$ nuclei, using the argument
outlined on p.127.

5.5. From the result of Problem 5.4, expressing the
dipole operator in $N = Z$ nuclei as $-e \sum_i \tau_z^{(i)} \underline{r}_i$,
show that no E1 transition is possible between
$T = 0$ states. (Consider specifically the isospin
component of the wavefunction of a proton and a
neutron outside closed shells.) Return to
Problem 5.2 to determine whether this new selection rule has relevance.

6. Nuclear reactions

INTRODUCTION

In the earlier chapters the decay of unstable ground
states was considered, and in the previous chapter the
decay of low-lying excited states. The scope is extended here to states of high enough excitation to
decay by particle-emission and also to consideration
in more detail of the setting up of the low-lying
states which decay by γ-emission. The experimental
technique is to bombard a suitable material (the
target) with a beam of nuclear projectiles and to observe the reaction products, or to select them for
observation using coincidence techniques. If the
projectiles are charged particles (p, d, α, ^{12}C, ^{16}O,
etc.) they can be produced in quantity in a well-
focused beam of well-defined energy from an accelerator,
e.g. a cyclotron or an electrostatic generator. If
the projectiles are neutrons, then they must first be
produced by a nuclear reaction, selected in energy as
well as possible, and collimated (rather than focused)
onto the target; the products from the second nuclear
reaction are then observed - against the background
produced by neutrons from the first reaction emitted
in directions other than that of the second target.
A copious source of neutrons is a reactor (see Chapter
7), especially for neutrons of low energy (the keV and
eV region); but for neutrons of energy a few hundred
keV and greater it is preferable to produce them
using a primary reaction induced by charged particles
from an accelerator.

The study of nuclear reactions has three basic
aims:

(1)   to locate levels in a nucleus;
(2)   to determine nuclear properties of levels;
(3)   to study reaction mechanisms, i.e. nuclear dy-
      namics as distinct from the (quasi-) static
      behaviour in (2).

Experimentally there is much overlap of the three
aims, especially (1) and (2), since the location of
levels and decay mechanisms can often be determined
in the same experiment. To locate levels we need to
be able to handle the kinematics of nuclear reactions;
to determine their properties we need to look at the
decay modes and measure their lifetimes. For (3),
it is intended to limit the discussion to a brief
description of the direct and compound nucleus reaction mechanisms.

THE KINEMATICS OF NUCLEAR REACTIONS

A nuclear reaction can be represented as

$$A + a \rightarrow B + b + Q,$$

where a is the incident projectile and A the target
nucleus, at rest in the laboratory frame; b is a light
particle emitted in the reaction and B the final
nucleus (the distinction between light and heavy final
product is one of convenience; it is not always possible to make it - an example being fission (see
Chapter 7)). The 'heat of reaction' Q, as defined
above, is positive if the kinetic energies of the
final products exceed that of the initiating projectile. The extra energy comes from the internal poten-

tial energies of the particles, and these are reflected in their masses. Elastic scattering is a particularly simple example of a nuclear reaction, having $Q = 0$. In general, the final nucleus need not be in its ground state; an increased state of excitation results in a reduced Q-value. From the kinematics of the reaction the Q-value is determined and so the state of excitation of B, provided the ground-state transition can be recognized.

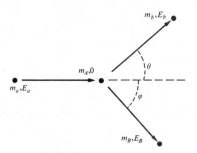

Fig.6.1. Schematic representation of the kinetics of the reaction A + a → B + b.

If the reaction is represented in the laboratory frame as in Fig.6.1, then using non-relativistic mechanics (accurate enough for most reactions using electrostatic generators and small cyclotrons):

conservation of momentum:

$$\sqrt{(2m_a E_a)} = \sqrt{(2m_b E_b)}\cos\theta + \sqrt{(2m_B E_B)}\cos\phi$$

$$0 = \sqrt{(2m_b E_b)}\sin\theta - \sqrt{(2m_B E_B)}\sin\phi \, ,$$

which on re-ordering, squaring, and adding give

$$m_B m_B = m_a E_a + m_b E_b - 2\sqrt{(m_a m_b E_a E_b)} \cos\theta ,$$

conservation of (kinetic) energy:

$$E_a + Q = E_b + E_B,$$

giving

$$(m_B + m_b)E_b - (m_B - m_a)E_a - 2\sqrt{(m_a m_b E_a E_b)} \cos\theta - m_B Q = 0,$$

(6.1)

conservation of (mass) energy:

$$m_A + m_a = m_B + m_b + Q/c^2 .$$ (6.2)

Note that in the non-relativistic limit it is convenient to express the relativistic conservation of
mass-energy as two separate equations connected by
the heat of reaction, for a reason which becomes apparent in the next paragraph.

In the general experiment, we measure the energy
of emission of b at a given angle of emission θ relative to the beam. In equations (6.1) and (6.2) it
is assumed that m_a, m_A, m_b, E_b, and θ are known (given
data or by measurement) and m_B and Q are unknown.
Thus (6.1) and (6.2) can be used to determine m_B and
Q. In point of fact it is rarely necessary to solve
these cumbersome simultaneous equations in the normal
way. From examination of the mass values of the
stable and nearly stable nuclei it will be seen that
the error incurred by using the mass number A for
the mass of an unknown nucleus (and A will be known by

conservation of nucleons) in (6.1) is probably of
the order of 0·1 per cent, and certainly not greater
than 0·3 per cent for all but the very lightest nuclei.
It is therefore possible to solve (6.1) directly for
Q to an accuracy $\sim$ 0·2 per cent; since in (6.2) the
term in Q is very small, the error incurred in deducing
an accurate mass value m_B from (6.2) using this inaccurate Q is probably less than the errors arising
from the quoted mass values of the other particles.
In any case, if it is justified to work to greater
accuracy relativistic effects should be considered.

If we have measured a number of different energies of particle b at a given θ, corresponding to
different energy levels in the final nucleus B (assuming that b itself has no excited state - true for
n, p, d, α), then to each energy E_i can be assigned
a Q-value Q_i, where $Q_0 - Q_i$ is the energy of the level
in nucleus B. Q_0, the Q-value of the transition to
ground, is either known or recognizable from a previously known pattern of excited states. Often the
largest Q-value is assumed to arise from the transition to ground - occasionally an inaccurate assumption.

Often equation (6.1) is used in reverse. In a
given spectrum we may suspect that there may be spur-
ious lines due to impurities in the target. We therefore wish to determine energies corresponding to
levels in the final nucleus following reaction with
the target impurity, for which Q_0 is known and the
energy levels of the final nucleus. Thus in (6.1)
all is known except E_b and this must be solved either
as a quadratic in $(E_b)^{\frac{1}{2}}$ or by iteration.

Finally, in measuring differential cross-sections,

it must be appreciated that the correct coordinate
system to use is the centre-of-mass frame, not the
laboratory frame. This affects the measurements in
that the transformation from the latter to the former
frame changes both the angle and the effective solid
angle of the detector. The calculation has been left
as an exercise.

PROPERTIES OF LEVELS

Having located (bound) levels in a nucleus using
the above kinematic analysis, it is desired to determine other characteristics of these levels such as
spin, parity, γ-decay spectrum to lower-lying levels,
half-life for γ-decay, magnetic moment, electric quadrupole moment, and less obvious characteristics such
as the extent to which a particular level looks like
a single particle (proton or neutron) outside a core
which could be the ground state (or an excited state)
of the appropriate neighbour.

A great deal of information can be obtained by
detecting the γ-radiation from the state, either
directly or in coincidence with the particle which
resulted in the formation of the state. The relative
intensities of γ-transitions to known levels can give
useful information on application of the selection
rules. By the method of 'delayed coincidences' it
is possible to plot the decay curve of the state and
so determine its half-life. Used in conjunction with
Weisskopf estimates (see p.125), it may then be possible to determine the spin and parity of the state.
A useful technique is to measure the angular correlation of the γ-radiation either relative to the beam,
with the outgoing particle detected in a certain direc-

tion (a most convenient direction is along, or against, the beam, since the problem then has axial symmetry).

In order to see how information can be obtained from the direction of emission of particles from the nucleus, consider an idealized nuclear state which looks like an α-particle in an $l = 1$ orbit outside a closed shell core. Then, relative to a given axis, the angular components of the wavefunction will be of the form $Y_{1,m}(\theta,\phi)$. If the reaction setting up the state was selective in m and ensured that it was in the $m = 0$ sub-state then the angular wavefunction is $Y_{1,0}(\theta,\phi) \propto \cos\theta$; if also this state is unstable to the emission of α-particles, then the wavefunction of the free α-particle must have the same angular dependence (a necessary condition for matching wavefunctions at a spherical boundary). So the probability of observing an α-particle at an angle θ to the given axis will in this simple case be proportional to $\cos^2\theta$. It is unfortunate that the example chosen is that of an unbound state when discussing the γ-decay of particle-stable energy levels, but the object was to avoid the additional complexities of spin when considering effects of orbital angular momentum and this is not possible for γ-transitions (nor nucleon emission). Angular dependence is, however, implicit in the discussion of γ-emission in the previous chapter, where it was indicated that γ-absorption by a ($J=0,M=0$) orbital state of a proton to form a $(1,0)$ orbital state arose from the z-component of the radiation field. Since the field is perpendicular to the propagation vector, the z-component will contain the factor $\sin\theta$, where θ is the angle between the propagation vector and the z-axis. Thus the probability

of absorption $\propto \sin^2 \theta$ and, by time-reversal, the probability of decay of a $(1,0)$ orbital state to a $(0,0)$
orbital state by emission of a photon in a direction
θ to the axis of quantization $\propto \sin^2 \theta$. Another way
of presenting this is merely to state the classical
result that a dipole along the z-axis emits a radiation
field whose amplitude is proportional to $\sin \theta$.

Thus, in general, if a level can be excited in
a single sub-state relative to a given axis then the
decay products (whether particle or γ-ray) will be
emitted with an angular variation relative to that
axis. A corollary to this statement is that if the level
has equal population of all sub-states then the radiation will be isotropic, because lack of knowledge concerning the formation of the state can only be expressed in this way. It is rarely possible to set
up a decaying level as a single sub-state, but neither
is it necessary to do so to observe an angular dependence of the decay; a non-random population of the sub-
states is sufficient, and this occurs whenever the
particle initiating the reaction is absorbed out of
a directional beam. Classically, the absorption of
a particle moving in a certain direction cannot change
the component of orbital angular momentum of the
system along that direction and this holds good in
making the transition to quantum mechanics. It is
therefore convenient to make the beam direction the
axis of quantization; the setting up of the state by
absorption of a particle out of the beam is then
characterized by $\Delta m = 0$. Thus if an α-particle were
to be absorbed by a 0^+ target to form a 1^- state then
this state could only be formed as the $(1,0)$ sub-state,
and its subsequent decay by γ-emission to a 0^+ ground

state will give an angular distribution $\propto \sin^2 \theta$. If
the target ground state has angular momentum J and the
α-absorption produces a state $J' = J + 1$ then the sub-
state (J,M) will result in the formation of $(J + 1,M)$.
Since nothing is known about the target condition the
angular distribution arising from each initial (J,M)
must be summed with equal weighting. But in the intermediate state, the sub-states $(J + 1, J + 1)$ and
$(J + 1, -J - 1)$ cannot be formed at all, so the averaging in the initial state is not equivalent to random
averaging in the intermediate state and is not likely
to produce isotropy. To proceed further requires a
detailed study of the mathematics of coupling of angular momentum in quantized systems; here it suffices
to state that an angular distribution can be calculated in terms of a few nuclear parameters, the
J^{π} of the states involved, the mixing of emitted
radiations (e.g. M1 and E2), etc., and checked a-
gainst measurement to give information concerning
these parameters.

Two important qualitative features can be appreciated. If the intermediate state has $J = 0$ then any
radiation subsequently emitted must be isotropic -
since randomness is represented by equal populations
of all sub-states, $J = 0$ must always be in a random
condition. Less obviously, if the particle absorbed
has $l = 0$, then all subsequent radiations will be
isotropic - this follows from the realization that
adding nothing (angular-momentum-wise) to randomness
can hardly produce other than randomness.

The above outlines the theme, but the variations
are numerous and can be complicated. As an example of
complication, if a reaction is observed in the presence
of a magnetic field then, from Larmor's theorem, the

whole system can be assumed to rotate with the Larmor precession frequency; measuring the angular distribution as a function of time after formation of the intermediate state should reveal a rotation of the angular distribution by an amount which depends upon the magnetic moment of the intermediate state and the time delay and so gives the magnetic moment of this excited state.

The methods for determining to what extent a given state looks like a shell-model state are best left until the next section.

INTERACTION MECHANISMS

Broadly speaking interactions can be divided into two categories, but the division is not sharp. The category of 'direct reactions' refers to reactions which take place whilst the bombarding particle is traversing the target nucleus, i.e. in a time $\sim 2R_N/v$, where v is the velocity of the particle (typically $> c/10$). This characteristic time is $10^{-21} - 10^{-22}$ s. An example is the reaction A(d,p)B, in which the deuteron, whilst passing through the periphery of nucleus A loses a neutron to form B. In the second category, 'compound-nucleus reactions', the time of interaction is very long compared with the characteristic time above. Thus there is a long time interval between the absorption of the bombarding particle and the emission of the product particle. It is therefore reasonable to talk of the compound nucleus as an intermediate product in the reaction; if the reaction is A(X,Y)B then it effectively proceeds in two steps $A+X \rightarrow C^* \rightarrow B + Y$. Since C^* exists for a time long on the nuclear time scale it must correspond to a (quasi-) stationary state of the nucleus C and is in fact just one of the many virtual states of excitation of C.

As is usual, we define the processes in terms
of extremes. The real situation is not always so
clear. If an incoming projectile interacts with a
nucleon near the surface of the nucleus, it or the
nucleon could be scattered clear of the nucleus giving
a fast direct reaction; if it gives up some energy
to the nucleon but insufficient to raise the latter
out of the nucleus and thereby moves further into
the nucleus, it can once again be scattered out by
another nucleon or share energy with it and still
remain inside the nucleus. At each collision it becomes harder for the particles concerned to escape,
and finally the energy is well shared among all the
nucleons. At this stage the compound-nuclear state
has indeed been formed. It can only decay when, by
chance, one nucleon can acquire a sufficiently large
share of the excess energy to become unbound, and
this can take a long time when the energy has been
shared over a large number of nucleons. Thus virtual
nuclear states tend to be longer lived (for a given
excitation) in heavier nuclei. At low energies, and
especially for charged particles, the compound-nucleus
process will dominate in the reaction cross-section
(the elastic Rutherford scattering is large and
direct) since, having penetrated the barrier (an improbable process), the particle even if it fails to
collide has little chance of penetrating once again
at the first pass. It is therefore reflected back
and forth at the surface until finally it shares
energy and forms the compound nucleus. At high energies direct reactions become of increasing importance
- at very high energies it is virtually certain that

both incoming and struck particle will escape from
the nucleus.

The compound-nucleus process

The reaction can be described mathematically as
arising from a first-order time-dependent perturbation
setting up the compound nucleus C* which subsequently
decays, or alternatively as arising from a second-order
time-dependent perturbation between the initial and
final states which is dominated by one channel proceeding through C*. The two processes are in reality
the same thing, but it is instructive to look at both.

As always in time-dependent perturbation, we
consider the amplitudes a,b,c of normalized wavefunctions of states A,B, and C* to be functions of
time. The wavefunction representing C* is taken to
be of the form $c_0 \exp(-t/2\tau)\psi_C(r,\theta,\phi)$ in order to represent the fact that, left alone, C* will decay in
intensity with mean life τ. First-order perturbation
theory gives the rate of formation of C* as

$$-(i/\hbar)H_{CA}\exp\{i(E_C-E_A)t/\hbar\}a,$$

where H_{CA} will be assumed to be independent of time,
except for having been switched on at $t = 0$. The amplitude c therefore satisfies the differential equation

$$\frac{dc}{dt} = -(1/2\tau)c - (i/\hbar)H_{CA}\exp\{i(E_C-E_A)t/\hbar\}a .$$

If the probability of transition is small, then there
is little absorption of the wave representing the
initial state, so a = constant = 1 is a reasonable

approximation, and the equation integrates to

$$c = \frac{-(i/\hbar)H_{CA}[\exp\{i(E_C-E_A)t/\hbar\}-\exp(-t/2\tau)]}{(i/\hbar)(E_C-E_A) + 1/2\tau} \quad , (6.3)$$

satisfying the condition $c = 0$ at $t = 0$. Whilst τ is large on a nuclear time scale it is small on a laboratory scale ($\sim 10^{-14}$s or less), and so the term $\exp(-t/2\tau)$ can be neglected under experimental conditions giving

$$|c|^2 = \frac{|H_{CA}|^2}{\{(E_C-E_A)^2+\hbar^2/4\tau^2\}} = \frac{|H_{CA}|^2}{\{(E_C-E_A)^2+\Gamma^2/4\}} \quad , (6.4)$$

where $\Gamma\tau = \hbar$. Now the probability of decay of a state is the sum of its decay probabilities to the various decay channels:

$$1/\tau = \sum_i (1/\tau_i) \quad \text{or} \quad \Gamma = \sum_i \Gamma_i .$$

Rate of formation of state B = $\dfrac{|c|^2}{\tau_\beta} = \dfrac{|H_{CA}|^2\Gamma_\beta}{\hbar\{(E_C-E_A)^2+\Gamma^2/4\}}.$

Considered as a second-order perturbation the rate of production of B is

$$(2\pi/\hbar)\left|\sum_c H_{CA}H_{BC}/(E_A-E_C)\right|^2 \times \text{(energy density of final states)}.$$

The time-dependence of the (decaying) state C* is

$$\exp(-iE_C t/\hbar)\exp(-t/2\tau) = \exp\{-i(E_C-i\Gamma/2)t/\hbar\}.$$

Thus the decaying state C* is effectively at an energy $(E_C - i\Gamma/2)$. If in the above summation one term

dominates and that one is C*, then the expression inside the moduli becomes

$$\frac{|H_{CA}|^2 |H_{BC}|^2}{\{(E_A-E_C)^2+\Gamma^2/4\}} .$$

Comparison of the two expressions for rate of formation then gives

$$\Gamma_B = 2\pi|H_{BC}|^2 \times \text{(energy density of final states)},$$

and, by applying a time-reversal argument,

$$\Gamma_A = 2\pi|H_{CA}|^2 \times \text{(energy density of final states of time-reversed system)}.$$

From Chapter 4 (p.101) this is

$$\Gamma_A = 2\pi|H_{CA}|^2 \{\frac{4\pi p_a^2}{v_a(2\pi\hbar)^3}\} (2J_a+1)(2J_A+1),$$

where the surrounding box has been taken to have unit
volume (in any case its size drops out), and the
statistical weights of the states have been included.
Thus

$$\Gamma_A = (1/\pi\hbar^3)|H_{CA}|^2 m_a^2 v_a (2J_a+1)(2J_A+1) , \qquad (6.5)$$

and the rate of formation of B is then

$$\frac{\pi\hbar^2(2J_C+1)}{m_a^2 v_a(2J_a+1)(2J_A+1)} \cdot \frac{\Gamma_A\Gamma_B}{(E_A-E_C)^2+\Gamma 2/4}$$

where the statistical weight of C* has been inserted
since the calculation has previously referred to a

single intermediate state. The statistical weights have been introduced rather too glibly. The factor $1/(2J_a+1)(2J_A+1)$ is perhaps obvious since the Γs have been defined in terms of a summation over final states, whereas the initial state has unit normalization spread over the sub-states, i.e. an average over initial states is required. In any case the argument presented is sound; the density of final states in the time-reversed system certainly contains the factor, but the yield does not and therefore must contain $\Gamma_A/(2J_a+1)(2J_A+1)$. At first sight it may appear that, at least in the expression given for the second-order transition, a term like $(2J_C + 1)^2$ arises, since the summation over sub-states is done before taking the square of the modulus. However, the expressions derived are for rates integrated over all directions of emission of the final particle and all cross-products integrate to zero, leaving only the terms which arise if the summation over sub-states were taken after forming the square of the modulus. The term $(2J_C+1)$ is the number of 'final' states produced by the first-order process; that these states decay is not relevant to the first-order treatment. Hence the right side of equantion (6.4) should be increased by this factor.

From Appendix A, the cross-section for a process is defined such that $\sigma_{AB} \, v_a$ gives the rate of formation of state B. Therefore

$$\sigma_{AB} = \frac{\pi\hbar^2}{m_a^2 v_a^2} \cdot \frac{(2J_C+1)}{(2J_a+1)(2J_A+1)} \cdot \frac{\Gamma_A\Gamma_B}{(E_A-E_C)^2+\Gamma^2/4},$$

$$(6.6)$$

and the factor in the first brackets can be written
more familiarly as $\pi \lambdabar^2$. In this expression E_A is the
energy of the incoming beam of particles (in the
centre-of-mass frame) and is therefore the energy
variable. The shape of the resonance curve is shown
in Fig.6.2, where the variation of $\pi \lambdabar^2$ over the resonance has been neglected.

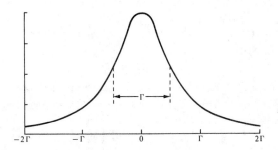

Fig.6.2. Shape of Breit-Wigner resonance denominator plotted
against $(E-E_r)$, with energy units of Γ. If the resonance is sharp so that neither $\lambdabar^2$ nor the widths
vary appreciably over an energy range $\sim 4\Gamma$, then this
curve will represent the behaviour of the reaction
cross-section.

In the derivation of the Breit-Wigner formula
above it has been assumed that the decay of C* is
characteristic of C* alone and does not depend upon
the way C* was produced. This independence of formation and decay of the compound nucleus was first
stressed by Bohr. It is an important point, but has
been somewhat overworked in the textbooks. It holds
only for an isolated resonance, but in tests quoted

different formation processes can populate the overlapping resonances to different degrees, and the
decay processes will then differ. If the number of
overlapping resonances is very large then a statistical averaging process will reduce the difference.
Thus independence of formation and decay is good for
a single resonance or a large number of overlapping
resonance, but probably does not hold for a few
overlapping resonances.

     Two outstanding examples of resonant reactions
are shown in Figs. 6.3 and 6.4. In the (p,γ) reaction
the resonances are sharp because the proton width is
small due to the effect of the potential barrier.
From the above, Γ is proportional to the square of

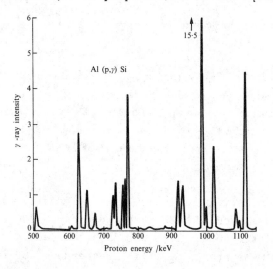

Fig.6.3. Resonances in ^{27}Al (p,γ) ^{28}Si. The resonance widths
         indicated represent experimental limitations; the true
         widths are considerably less. (Based on BROSTROM
         et al. (1947). *Phys.Rev.* <u>71</u>, 661.)

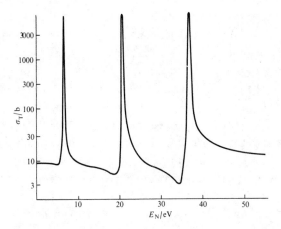

Fig.6.4. Resonances in ^{238}U(n,γ)^{239}U for neutrons up to 50 eV.
Note the large cross-sections at resonance, the high
density of levels in the compound nucleus ($\sim$ 1 every
15 eV) and the narrow widths. Note also that the
measurements were made in absorption. Off-peak the
absorption is due to 'hard-sphere' scattering inter-
fering with the scattering from tails of the resonances;
on-peak the absorption is dominated by the capture process.

the modulus of a matrix element which connects a
state representing a particle within the nucleus with
a state representing the same particle outside the
nucleus (a plane wave at infinity) and the overlap of
this latter wavefunction with the nucleus introduces
the barrier-penetrability term. It is then surprising
that the neutron resonances shown have even smaller
neutron widths. This is partly accounted for by the
fact that the separation between resonances has become
small because in the example shown the target nucleus
in (n,γ) is a heavy one and in (p,γ) a light one. But

why should crowding of levels result in smaller partial
widths? This question can be conveniently answered
by considering what would happen to cross-sections
were it not the case. Assume that the cross-section
is being averaged over an energy range ΔE large compared with resonance widths, then the contribution
to the *total* reaction cross-section of a single resonance is obtained by integrating the Breit-Wigner
expression (with Γ for Γ_B) giving a result proportional to Γ_A only. The number of levels in ΔE is
$\sim \Delta E/D$, where D is the average spacing between levels.
The averaged cross-section is therefore $\propto \Gamma_A/D$. As
the energy is increased D will get smaller corresponding to an increase in level density with excitation of nucleus C. If Γ_A remains approximately
constant with energy then the cross-section increases
without limit. But cross-sections at high energies
should approach something like the classical πR_N^2,
and we are forced to conclude that (Γ_A/D) is roughly
independent of energy. Thus the partial width into
a given channel depends upon the local spacing of
levels (see also Chapter 3). This does not mean that
the total width of the resonance is similarly limited;
it is a sum of all partial widths and each such partial width is proportional to the local spacing, but,
as the energy increases, the number of open channels
contributing also increases. Thus the total width of
levels tends to increase with energy, but only slowly
since the two effects almost compensate.

To return to (n,γ), it was stated that the above
considerations only partly account for the small
neutron widths. A closer look at the expression for
the partial width in terms of the matrix element

(p.147) reveals that the width is proportional to the
velocity in the channel and this is very small for
neutrons of a few eV - so small that in this region
$\Gamma_n < \Gamma_\gamma$. (This surprising circumstance is also a consequence of the fact that γ-radiation is taking place
from a level at an excitation of $\sim$ 6 MeV or more in
a large nucleus, so there is an extremely large number of available states of lower energy to which γ-
transitions can take place.) In addition, the width
includes the penetrability through the angular-momentum
barrier (and for charged particles it is augmented
by the Coulomb barrier) with the result that Γ_n in
the eV region is far too small, when $l \neq 0$, to give
observable resonances.

In the case of the (p,γ) resonances, their sharp-
ness is determined largely by the barrier factor.
The 'widths without barrier' in light nuclei are
quite large fractions ($\sim$ a few per cent) of the single-
particle proton width, but the barrier penetrability
reduces them to a few eV giving a total width for
the resonance less than the experimental resolution
of the beam. The peak yield is not therefore the
yield at resonance, but a measure of the integrated
yield over the resonance.

Finally, the Breit-Wigner formula can be used
to show that neutron cross-sections vary as $1/v$ at
very low energies. So far it has been assumed that
the Γs are constants in the formula but this is an
oversimplification; the matrix elements can have an
inbuilt energy-dependence and, as has been mentioned,
there is a factor v_a in the expression for Γ_a. For
a sharp resonance at high energy in the incident
channel the variation is not expected to be large over

the region of large cross-section, but at very low energies the $\pi \lambda^2$ term becomes large and enhances the low energy tail of the resonance. It is therefore necessary to consider the contribution made by a resonance far from the peak of the resonance denominator. The cross-section is then of the form $\sigma \propto \pi \lambda^2 \Gamma_n \Gamma_\gamma / (E_C)^2$, assuming that $E \ll E_C$ and also $\Gamma \ll E_C$. Now Γ_γ is unlikely to vary whilst the energy changes by a few eV since the level in C is 6 MeV or higher in excitation, so the energy variation is contained in $\lambda^2 \Gamma_n$, and since $\lambda \propto 1/v_n$ and $\Gamma_n \propto v_n$ then $\sigma \propto 1/v_n$.

Direct reactions

The characteristic of a compound-nucleus resonance is a cross-section which changes rapidly with energy over a small range of energy. The range over which the cross-section changes rapidly is of order Γ and is correlated with the time scale of the process by $\Gamma \tau \sim \hbar$. A direct reaction has been defined as one taking place whilst the bombarding particle passes through the nucleus and therefore in a characteristic time $\sim 10^{-22}$ s. Using the above relationship, it is associated with a width ~ 10 MeV. Thus a characteristic of direct reactions is that the cross-section should not change rapidly with energy, or rather should not exhibit sharp maxima, since it is possible for a direct reaction involving charged particles below the barrier to increase exponentially in cross-section.

Another characteristic concerns the angular distribution of the process. If a reaction proceeds via a single (and this must be emphasized), well-defined resonance then, at the intermediate stage of the process it has a well-defined parity (the parity of the state C*). In setting up the state a definite direction has

been singled out (the direction of the beam), but
this does not correspond to a well-defined parity, so
in forming C* the information concerning direction
of motion is lost but not the axis of the beam. (To
form a state of well-defined parity it is necessary
to superimpose upon the state defined by the beam, .a
corresponding state with the beam moving in the opposite direction.) Thus the subsequent decay of C* will
give an angular distribution having symmetry about 90°
(fore-aft symmetry). The direct reaction has no such restriction, the symmetry axis contains the information concerning direction and angular distributions do not
in general have fore-aft symmetry. In fact there is
a strong tendency to forward-peaking, an extreme example being the case of absorption of a fairly high-
energy neutron beam which results in the diffractive
scattering of the beam. As in the classical case of
diffraction of light, the forward-directed scattering
peak has angular width $\sim \lambda/R_N$ rad. Many other types
of direct reaction conform to this simple diffractive
picture.

The stripping reaction is an interesting example
of a direct reaction. It is the reaction A(d,p)B (or
equivalently (d,n)), in which one nucleon of the incoming deuteron interacts with the nucleus whilst the
other passes it by. The probability of the process
occurring depends upon the square of the modulus of
$\int \psi_f^* V \psi_i \, d\tau$, where the initial wavefunction ψ_i is that
of the deuteron in the form of a plane wave combined
with its internal function, multiplied by the target
wavefunction ψ_A, whilst the final wavefunction ψ_f
will be that of the proton in the form of a plane wave
multiplied by the wavefunction of the final nuclear

state ψ_B, which for simplicity will be taken as $\psi_A \psi_n$, where ψ_n is a single-particle wavefunction of the neutron about the core A. Notice that the Coulomb distortion of the plane waves has been neglected here, though it is included in the analysis of research data. For the perturbation V we can put in the inter- action between the neutron and the nucleus A, which is taken to be of the form $V(r_n)$ with the centre of A defined as the centre of the coordinate system. It is also assumed that any change of shape of the deut- eron during its journey past the nucleus can be ne- glected and that reaction only occurs when the proton and neutron in the deuteron are at their common centre. Another simplifying assumption made here is that the reaction takes place only at the surface of nucleus A - this is not assumed in a more serious calculation.

With these approximations, in the integral, the internal wave-function of the deuteron becomes merely a constant (the probability that the proton and neutron should be found at its centre), and its plane wave be- comes $\exp(i\underline{k}_d \cdot \underline{r}_n)$, whilst that of the proton becomes $\exp(i\underline{k}_p \cdot \underline{r}_n)$, since $\underline{r}_n = \underline{r}_p = \underline{r}_d$, by assumption. By the same token $V = V(R_N) = $ constant, as also is $\psi_A^* \psi_A$. In the integral $\underline{r}_n \equiv (R_N, \theta', \phi')$ and the integral becomes

$$\text{constant} \times \int \psi_n^*(\underline{r}_n) \exp(i\underline{k}_d \cdot \underline{r}_n) \exp(-i\underline{k}_p \cdot \underline{r}_n) d\Omega'.$$

If $\underline{k}_d - \underline{k}_p = \underline{q}$, and the direction of $\underline{q}$ is taken to define the z'-axis the integral $\propto \int Y_{l,m}^*(\theta', \phi') \exp(iqR_N \cos\theta') d\Omega'$, where the single- particle angular wavefunction has been inserted for the neutron, but its spin function has been neglected.

The integral over ϕ' requires that $m = 0$, giving

$$\int_{\cos \theta' = -1}^{\cos \theta' = +1} Y_{l,0}(\theta') \exp(iqR_N \cos \theta') d(\cos \theta').$$

To see how this behaves take the cases $l = 0$ and $l = 1$. The first integral, for $l = 0$, can be performed at sight, giving constant $\times$ ($\sin qR_N/qR_N$), which, when the substitution $q^2 = (k_d - k_p)^2 + 4k_d k_p \sin(\theta/2)$ is made, gives a maximum at $\theta = 0$, followed by subsidiary maxima of reduced intensity. (The function $(\sin qR_N/qR_N)^2$ is recognizable as the diffraction function, so, provided $|\underline{k}_d - \underline{k}_p|R_N$ is less than π, the statement made is true. This is the case in most experimental applications.) Putting in $Y_{1,0} \propto \cos \theta$, for $l = 1$, gives an integral $\int_1 \exp(iqR_N c)c dc$. Integrating by parts gives $(\sin qR_N - qR_N \cos qR_N)/(qR_N)^2$ which, on squaring, is seen to give zero at $q = 0$, rising to a large first maximum, and then having subsequent lesser maxima. With the same proviso as for $l = 0$, the angular distribution will be non-zero at $\theta = 0$ (unless $k_d = k_p$ which is possible, for low E_d and a high Q-value, but not likely) rising to a large first maximum (at a fairly small angle in practical cases) followed by subsidiary maxima. The important point is that the location of the primary maximum can be used to determine the l-value of the transfer, i.e. the amount of orbital angular momentum taken in by the stripped nucleon. Thus the parity of the final state is known immediately on analysis of the angular distribution of the particle not captured, and the range of J-value of the final state is limited. The degree of limitation depends upon J for the initial nucleus - if this

were zero then the final state must have $J_f = l_n \pm \frac{1}{2}$,
where l_n is the transferred angular momentum. But in
the general case J_f can take the range $(J_i + l_n + \frac{1}{2})$
down to the smaller of $|J_i - l_n \pm \frac{1}{2}|$. Examples of
angular distributions in (d,p) are given in Fig.6.5.

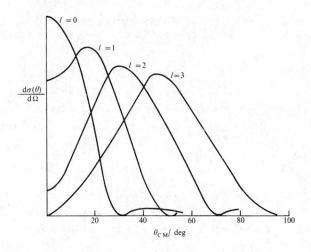

Fig.6.5. Typical angular distributions for the (d,p) direct reaction. If each process represented the full single-
particle transfer, then relative to $l = 0$, the curves
for $l = 1,2,3$ should be multiplied respectively by
$\frac{1}{4}$, $\frac{1}{12}$, $\frac{1}{30}$.

A useful feature of the stripping reaction is
that the magnitude of the cross-section is a measure
of the degree to which the final state looks like the
original state plus a single particle of the type indicated by the angular distribution. Of especial importance is the fact that the final state may be
bound to the particle transferred (indeed the calcula-

tion is simpler for such bound states, and the simple
assumptions made here would not be good for an unbound state). Because the transfer is virtual rather
than real, the real positive momentum q can be associated with a negative total energy of transfer.

The stripping reaction is a typical direct reaction, of which there are also many other examples.
The inverse (p,d) or pick-up reaction, (^{3}He, d), (α,p),
etc., and their inverses are further examples of
stripping. Other direct reactions are elastic and inelastic scattering of charged particles (usually α-
particles) off distorted nuclei; Coulomb excitation -
the electromagnetic excitation of a nucleus by the
rapidly changing field of a passing charged particle;
transfer reactions, e.g. (^{17}O, ^{16}O), in which a
neutron is transferred from the projectile to the
target - this is really an example of stripping with
heavy-ion projectiles.

PROBLEMS

6.1. For elastic scattering of a projectile of mass
m by a target of mass M show that the angle of
scattering in the laboratory system (θ) and
the centre-of-mass system (ϕ) are connected by
$\phi = \theta + \sin^{-1}\{(m/M) \sin \theta\}$ and that the ratio
of solid angles between the two systems is given
by $d\Omega_{CM}^{(\phi)}/d\Omega_{lab}^{(\theta)} = \sin^2\phi/\sin^2\theta \cos (\phi-\theta)$. Explain the significance of these transformations
in the measurement of differential cross-sections.

6.2*. A foil of $^7_3$Li is bombarded with protons of
5·0 MeV, and it is found that at 90° neutrons of
2·3 MeV are emitted. If the atomic masses of p,

n, and ^{7}Li are respectively 1·0078U, 1·0087U, and 7·0160U find the atomic number and mass of the daughter nucleus.

6.3*.A thin target of ^{27}Al is bombarded with 8 MeV protons, and the energy spectrum of the protons scattered at 45° to the incident beam contains strong peaks at 7·84 MeV and 6·99 MeV. What can be deduced about the levels of this nucleus?

6.4. By using an argument involving parity, show that $J = \frac{1}{2}$ for the intermediate state in a γ-γ cascade will, like $J = 0$, result in an isotropic angular correlation. Show also that the γ-decay of an isomeric $J = \frac{1}{2}$ state will be isotropic even when the state has been completely polarized into one of its sub-states (e.g. by application of a magnetic field at low temperature), but will not be isotropic in this latter condition if the γ-detector is sensitive to the state of circular polarization.

6.5*.In a given nucleus, a state C (0^+) decays by γ-cascade through a state B to the ground state A (0^+). If the two γ-transitions are E1, what is the spin and parity of B? Will there be a non-isotropic angular correlation of the two γ-rays? Alternatively, if the γ-radiations are again E1 but B is known to be 0^+, what are the spins and parities of A and C? Will there be an angular correlation?

For the more general case, discuss qualitatively how angular-correlation measurements can help to give information about spins and parities of nuclear states (a) if the multipole order of each γ-ray is known (e.g. from internal conversion measurements), (b) if it is not.

6.6*. Suppose that the reaction $^{12}C(\alpha,\gamma)^{16}O$ (Q = 7·15 MeV) shows an isolated resonance at E_α = 10·10 MeV with Γ_{tot} = 0·20 MeV, Γ_α = 0·15 MeV, and Γ_γ = 10 eV. The spin of the compound state is J = 4. The only other decay mode of ^{16}O at this excitation is by proton emission to the ground state ($J = \frac{1}{2}$) of ^{15}N. When the same level is excited by $^{15}N(p,\gamma)^{16}O$ (Q = 12·11 MeV) what will the peak cross-section be? (Remember to consider centre-of-mass effects.)

6.7*. (a) Write an equation embodying the principle of detailed balance as it applies to two-body reactions with unpolarized beams and targets. (This should be deduced from Fermi's golden rule and the definition of cross-section - since these have been used to deduce the Breit-Wigner formula it can also be conveniently extracted from the latter, though it has greater generality.) Explain how it may be used to deduce the spin of the pion from the measured cross-sections for $p + p \rightleftharpoons d + \pi^+$. (Note there is in addition a factor arising from the production of two identical particles when the reaction proceeds from right to left.) (b) Explain why the Breit-Wigner formula is important in the theory of slow neutron capture by nuclei. Show that, at a resonance peak (neglecting any background), $\pi\sigma^2 = \lambda^2 g\sigma_s$, where σ and σ_s are the total and elastic scattering cross-sections for neutrons.

7. Fission and Fusion

In these days of energy crises no textbook on nuclear physics would be complete without some reference to these two sources of energy. Fusion has very little overlap with nuclear physics, except for the fact that the basic reactions are between light nuclei, but the nuclear reactor based on fission can truly be said to be the brain-child of the nuclear physicist. Indeed virtually the whole of nuclear physics is involved in describing how and why a reactor works.

THE FISSION PROCESS

In Chapter 1, the mass formula indicated that all nuclei above $A \sim 150$ are in principle unstable to α-emission. At the same time, the question of stability against splitting up in any way whatsoever might have been considered, but was not. It is so now. Splitting into two approximately equal parts is expected to be possible at large enough mass values, since the mass formula indicates that medium-weight nuclei are most stable. Expressing $M(Z,A) - M(Z_1,A_1) - M(Z_2,A_2)$, where $Z = Z_1 + Z_2$, $A = A_1 + A_2$, in terms of the mass formula (p. 18), and maximizing this function, gives the result $Z_1 = Z_2$, $A_1 = A_2$, and

$$M(Z,A) - 2M(\tfrac{1}{2}Z,\tfrac{1}{2}A) = -\beta A^{2/3}(2^{1/3}-1) + \varepsilon Z^2 A^{-1/3}(1-2^{-2/3}),$$

neglecting the trivial term in δ. The condition for stability is therefore

$$\frac{z^2}{A} < \frac{\beta}{\varepsilon}\left(\frac{2^{1/3}-1}{1-2^{-2/3}}\right) \sim 0\cdot7(\frac{\beta}{\varepsilon}) \sim 17 \ .$$

Thus, in principle, all nuclei above $A = 90$ ($Z = 40$) are unstable to fission into equal fragments. This low value is reminiscent of the low value obtained for α-decay, which in turn leads to the reason why such light elements do not fragment, namely, the Coulomb barrier. It might be more realistic to expect fission to occur if the mass of the nucleus under consideration exceeds the masses of the two parts by more than the Coulomb energy when the two parts have just separated. Fission will then occur over the top of its barrier. This adds a term

$$\frac{(Ze/2)^2}{4\pi\varepsilon_0\cdot2R_0(A/2)^{1/3}}$$

to the energy of the two fragments. Inserting a suitable value for $R_0(1\cdot3$ fm), or alternatively assuming that ε derives from a uniform charge distribution and using the value of ε to determine the radius, modifies the condition to $Z^2/A > 50$ for instability, and this would correspond to an extrapolation to $Z \sim 130$ along the valley of stability.

These estimates are very crude. It has been assumed that, in inverse, as they move toward each other the two fragments remain spherical up to contact. But nuclei in isolation can be non-spherical, and even if spherical it is not realistic to assume they are rigid. In the presence of each other it is possible that the Coulomb potential be less than the monopole term predicts because of induced distortion or because the proton distribution inside

each nucleus may no longer be uniform. These and
other mechanisms can diminish the Coulomb potential
near the point of contact whilst not greatly affecting
the potential of the single large nucleus. It is to
be expected that spontaneous fission will occur for
less massive nuclei than predicted by the above formula.

To consider the mechanism in more detail, a potential function is needed to describe the interaction between the two fragments. When the two components are distinct, the potential can be expressed
as a function of the distance between their centres,
but this parameter has little meaning before separation. In this region a convenient parameter is ΔR
for an ellipsoidal nucleus as defined in Chapter 1
(p.31), but this ceases to be useful near the point
of break-up since $\Delta R \rightarrow \infty$ produces a needle-like
nucleus rather than two globules. If we ignore this
difficulty the potential is shown in Fig.7.1, where
the two dotted curves correspond to the limiting cases
of a nucleus in principle unstable to fission but
not observed to decay so and of a nucleus which would
decay by the fission process in a time of order of
the nuclear characteristic time. The interesting case
is shown by the full line which represents a typical
distorted nucleus in the region of uranium.

If the well representing the large nucleus is
extremely shallow then it is possible for states inside it to decay by penetrating the fission barrier
(since the particle penetrating has large effective
mass, the penetrability varies extremely rapidly with
energy below the top of the barrier so that shallow
means of the order of a few hundred keV). This leads

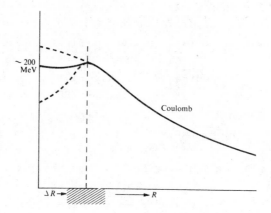

Fig.7.1. Schematic fission potentials. Note that the abscissa
         is the distortion parameter ΔR up to somewhere below
         separation and the distance between fragment centres
         above separation. The dashed curves are explained
         in the text.

to a class of radioactivity not previously discussed,
namely, decay by spontaneous fission. If the well
is deeper than this, then the levels near ground may
be stable to this mode of decay, but levels of higher
excitation will not be, so fission will take place
in preference to γ-emission. Such levels could be
excited by inelastic scattering, e.g. (n,n') or (p,p').
At a greater well-depth, the location of the top of
the fission barrier relative to neutron instability
is of great importance. If the fission barrier
lies lower then the nucleus one neutron less than
the one here considered, on absorbing a thermal neutron, will decay by fission; if the fission barrier

lies say 1 MeV higher then neutrons of energy greater
than 1 MeV will be necessary to induce fission.
Notice that a neutron threshold is well defined, but
the fission threshold is effective in that just below
it the process can take place but with very much reduced probability.

As an example of the above, spontaneous fission
occurs in the following:

$$^{235}_{92}U \ (\tau_f \sim 2 \times 10^{17} \text{years}), \quad ^{238}_{92}U \ (\tau_f \sim 10^{16} \text{years},$$

$$^{240}_{94}Pu \ (\tau_f \sim 10^{11} \text{years}), \quad ^{250}_{98}Cf \ (\tau_f \sim 10^{4} \text{years}),$$

$$^{254}_{98}Cf \ (\tau_f \sim 55 \text{days}).$$

Nuclei which undergo fission on capture of thermal
neutrons are ^{233}U, ^{235}U, and ^{239}Pu, whilst nuclei
which require capture of fast neutrons are, with
effective neutron thresholds in brackets,
$^{232}_{90}Th$ (1·3 MeV), ^{234}U (0·4 MeV), ^{236}U (0·8 MeV),
^{238}U (1·1 MeV), and $^{237}_{93}Np$ (0·4 MeV). The two last
sequences reflect the importance of even or odd
neutron numbers, emphasizing the effect of pairing
of nucleons.

For nuclei in this region the fission reaction
is highly exothermic with a Q-value of order 200 MeV.
This is highly desirable for production of energy,
but of itself does not lead to power production. The
reaction must be self-sustaining as in chemical com-
bustion when part of the energy liberated in the reaction is used to heat the fuel and so encourage
further reaction. From information now to be dis-

closed, the fission reaction can be made self-sustaining. Fission is not merely the splitting of a
large nucleus into two smaller ones but, as is to be
expected, a lot of debris is produced at the same
time. Neutrons form part of this debris, either produced simultaneously with the large fragments or more
probably by secondary emission from the fragments
which will in general be formed in states of high
excitation. In addition, the fragments lie on the
neutron-rich side of the valley of stability (since
N/Z increases with Z along the valley) and therefore
have low binding energies for neutrons. These neutrons will be emitted almost instantaneously
($\tau \sim 10^{-16}$s or less) with a spectrum peaked at ~ 1 MeV.
This latter point can be understood in terms of the
discussion on widths in Chapter 6; Γ_n from a given
state to any final nuclear state is proportional
to k or $E_n^{\frac{1}{2}}$, but the density of final nuclear states
has very roughly an exponential dependence upon
energy of excitation. The most probable energy of
emission of a neutron is therefore the maximum of
$(E-E^*)^{\frac{1}{2}}\exp(\alpha E^*)$, where E is the energy of the neutron
transition to ground and E^* the excitation of the
final nucleus. At maximum $E-E^* = 1/2\alpha$. Peaking at
$E_n = E-E^* \sim 0\cdot 7$ MeV gives $\alpha = 0\cdot 7$ in units of $(\text{MeV})^{-1}$,
indicating how the level density behaves for a heavy
nucleus. Note that exp (αE^*) is an oversimplification, but it does represent roughly the crowding together of levels at higher energies; α varies slowly
from nucleus to nucleus.

In addition to these 'prompt' neutrons there are
also emitted some delayed neutrons amounting to
$\sim 0\cdot 5$ per cent of all neutrons produced. These were

initially a source of mystery, but are now known to
occur following β-emission; an example is the decay
of the fission fragment $^{87}_{35}$Br which β-decays with 55s
half-life to $^{87}_{36}$Kr, with an available energy $\sim$ 6 MeV.
As well as decaying to nucleon stable levels in ^{87}Kr,
a fraction of the β-transitions are to states of excitation greater than 5·4 MeV, and these can decay
by neutron-emission. Neutron-emission will therefore
be delayed by the time taken for β-decay of ^{87}Br.
Although relatively few, these delayed neutrons play
a large part in the control of a reactor. If each
neutron-induced fission event produces on average
more than one neutron, the possibility of a self-sustaining chain reaction arises. Typically the number
ν, per fission (average) is something like 2-2·5.

Before going on to discuss how to make use of
the fission reaction, comment should be made on the
reference to the fragment A = 87. This is a long way
from the symmetric splitting ($A \sim$ 115) which has been
suggested as giving the greatest energy release. In
fact the most probable mode of fission is not the
symmetric mode, but corresponds roughly to an A = 90
and 140 split. The explanation of this is still required; magic numbers probably play a part since
N = 50 for $A \sim$ 90, N = 82 for $A \sim$ 140. Note that the
Coulomb term is less for the asymmetric split relative to the symmetric, offsetting to some extent the
mass balance which favours the symmetric split. Pre-
diction of the 'mass spectrum' requires detailed
knowledge of the fission process - any mechanism
which can change the fission barrier height, as a
function of the fragment masses, by $\sim$ 0·1 MeV in
200 MeV could account for the spectrum. It is inter-

esting to note that, close to threshold, the (90, 140)
splitting predominates and there is virtually no
symmetric splitting. As the energy of the neutron in-
ducing fission rises, the splitting becomes more
nearly symmetric in accordance with the rough guide-
lines above.

THE CHAIN REACTION AND REACTORS

If a material fissile to fast neutrons produces
ν neutrons per fission (on average) and the probability
that, on absorbing a neutron (and the definition can
be extended to include absorption in other materials
present), fission occurs is p (again on average) then,
if $p\nu > 1$, a chain reaction will occur in a large mass
of this material, culminating in an explosion. The
intensity of the explosion depends on the rate of liberation of energy compared with the rate at which energy can be transmitted away (possibly taking the
material with it). If the interaction length for
fission is λ then the interaction time is $\lambda/\bar{v}$, and
in that (mean) time the neutron multiplication is
$(p\nu-1)$, giving an exponential growth in neutron flux
with time constant $\lambda/\bar{v}(p\nu-1)$. For ^{235}U, $\sigma_f \sim 1b$,
giving an interaction length ~ 20 cm, $\bar{v} \sim 10^9$cm s^{-1}
($E_n \sim 1$ MeV), and for $p\nu-1 \sim 1$ this gives a time constant $\sim 10^{-8}$s. For such a rapid rate of neutron
growth and therefore energy production, heat dissipa-
tion is negligible, and because of inertia the material
has no time to get away. The result is the so-called
'atom bomb'. From what has been said the critical
size for explosion is something greater than 20 cm,
and this can be achieved by imploding (chemically)
some smaller blocks of ^{235}U.

On the face of it, it seems unlikely that such
a system could be controlled. The fissile material
could be diluted and $p\nu$ could be made very close to
unity, but it would be difficult to produce just the
correct conditions for a slow enough growth rate in
order to be able to introduce some form of mechanical
control, and if by error conditions are changed to
make $p\nu$ slightly greater then the system runs amok.
At this point the fact that $\sim 0 \cdot 5$ per cent of the
neutrons are delayed becomes crucial; if $p\nu$ is just
less than unity for prompt neutrons and just greater
than unity for prompt-plus-delayed neutrons then the
time constant of the delayed neutrons takes control
and the growth-rate time constant will be considerably
greater than this (since $p\nu - 1 \sim 0 \cdot 005$). Thus there is
a range of variation of $p\nu$ of $\sim 0 \cdot 5$ per cent over which
the neutron flux is increasing at a controllable rate.
For the fast reactor, which is effectively being discussed here, control would be achieved by moving
parts of the fissile material away from the rest or
by moving some of the surrounding material relative
to the fissile core, thereby altering the number of
neutrons being reflected back into the core.

Historically the thermal reactor preceded the
fast reactor. Why was this the case? The answer
lies in understanding what makes a suitable fuel.
Although ^{238}U and some other nuclei listed on p.
167 are fissionable above a neutron threshold they
are not suitable fuels for a fast reactor, whereas
nuclei like ^{235}U, which are fissionable to thermal
neutrons, are. For these latter nuclei the fac-
tor p depends only upon the ratio $\sigma(n,\gamma)/\sigma(n,f)$,
since $\sigma(n,n')$, which is larger than $\sigma(n,f)$ at MeV

energies, does not remove neutrons but merely changes
their energies. However, if ^{238}U were used as a fuel
then (n,n') would rapidly remove neutrons below the
fission threshold and so be akin to absorption, with
a reduction in p. This process is enhanced by the
fact that the reactor has to have a structure and a
coolant so the neutron spectrum inside a fast reactor
is degraded to lower energies very quickly. It is
therefore necessary that the fuel be fissile down to
low energies. Thus natural uranium, containing only
0·7 per cent ^{235}U, cannot be made to go critical;
the fuel must be highly enriched in ^{235}U, a costly
undertaking.

It would therefore seem surprising that any
structure containing natural uranium can be made to
go critical. If, however, we look at the various
cross-sections at thermal energies ($\sim$ 0·025 eV) we
find that, for ^{235}U, $\sigma(n,\gamma) \sim$ 101 b and $\sigma(n,f) \sim$ 577 b,
whereas for ^{238}U, $\sigma(n,\gamma) \sim$ 2·76 and $\sigma(n,f) \sim$ 0. Even
taking into account that there is 140 times more ^{238}U
than ^{235}U in natural uranium, a value $p \sim$ 0·54 is obtained for thermal neutrons. Since $\nu \sim$ 2·44 for ^{235}U,
it is possible to obtain criticality provided that not
more than a quarter of the neutrons are lost in ther-
malizing them. Before considering the thermalization
process it is pertinent to ask why $\sigma(n,\gamma)$ for ^{238}U
turns out to be so much smaller at thermal energies
than the ^{235}U cross-sections. This is partly a for-
tunate accident, but not completely so.

On comparing the excitation functions of U + n
for the different isotopes, an odd-even effect is
immediately apparent. The even isotopes produce
narrow, well-separated resonances having very high

peak cross-sections; the odd isotopes produce broader
resonances, poorly separated (the density of resonances is ~ 10 times greater for the odd than the
even isotopes) and with lower peak cross-sections.
Previously it has been stated that level density increases exponentially with excitation, and the odd-
even effect does ensure that the excitations produced
by absorption in an odd isotope are about 1·5 MeV
higher than in an even isotope reflecting the fact
that in the former case the resulting ground state
has one more pair of neutrons than the initial ground
state ($^{235}U + n_{th}$ gives 6·48 MeV excitation, $^{238}U + n_{th}$
gives 4·78 MeV). But, because pairs must be split
in order to create excited states there is an odd-
even effect needed in applying the exponential function. It is probably more correct to translate
energies such that neutron thresholds coincide in
comparing level densities of neighbouring isotopes.
There is, however, a $(2J+1)$ factor appearing in the
level-density function, and this can account for the
observed ratio, since $^{235}U + n$ produces states of
$J^\pi = 3^-$ and 4^- (^{235}U ground state is $\frac{7}{2}^-$ and $l_n = 0$
for neutrons in the eV range) whilst $^{238}U + n$ pro-
duces $J^\pi = \frac{1}{2}^+$ only. Merely increasing the level-
density ratio should not (on average) enhance the
capture cross-section ratio at thermal energies,
since $\sigma_{n\gamma} \propto \Gamma_n \Gamma_\gamma / E_r^2$ (assuming that $E_r > \Gamma$ - see the
Breit-Wigner expression, p.148), and this can be
written

$$\frac{(\Gamma_n/D)(\Gamma_\gamma/D)}{(E_r/D)^2}$$

Provided the top two factors remain much the same

from one isotope to the other it would appear that
the thermal cross-section depends upon how close is
the nearest level compared to the level spacing - and
this is how chance creeps in. However, though
(Γ_n/D) is not expected to vary much between neighbouring nuclei since it refers to a single channel;
the same is not true for (Γ_γ/D), because it represents
a summation over a large number of γ-transitions. The
same $(2J+1)$ factor, which produced a greater density
of resonances in $^{235}U + n$, also provides a greater
density of states to which γ-transitions can be made
since strong γ-transitions change J only by one unit
at most. It therefore turns out that total radiation
widths Γ_γ, following neutron absorption, tend to be
similar for neighbouring nuclei, rather than Γ_γ/D.
There is therefore a tendency for even-A isotopes to
have lower thermal-capture cross-sections than odd-A
isotopes, apart from any odd-even effects which may
also occur because of pairing. This argument is
little affected by the presence or absence of fission
width, though it may need slight modification if the
increased total width no longer satisfies $E_r > \Gamma$.

Returning to thermalization, neutrons always lose
energy in elastic collisions, and the lighter the
scattering nucleus the greater the loss. Hydrogen
would therefore seem to be the most suitable material,
conveniently perhaps in the form of water; but, unfortunately it has a relatively high thermal-neutron
capture cross-section (~ 330 mb), and so absorbs many
of the neutrons on slowing them down. It cannot be
used with natural uranium, but uranium enriched in
^{235}U can go critical in a light (ordinary) water environment, and modern power reactors have been con-

structed to this basic design. The next suitable
material is heavy water since ^{2}H is the next best
thermalizer, and, moreover, $\sigma(n,\gamma)$ is very low (0·5 mb).
Again, modern power reactors have been built using
natural uranium fuel and heavy water. The next most
convenient material is carbon, which is much poorer
at moderating but still has a small enough $\sigma(n,\gamma)$ to
be of value (3·4 mb). The early experimental reactors
were built with carbon (graphite) as moderator and
natural fuel, and modern power reactors have been so
based.

The geometrical arrangement of fuel and moderator
is important. The obvious homogeneous mixture is unsuitable for natural fuel since the neutron capture
process in ^{238}U is important right down to its lowest
resonance at $E_n \sim 6\cdot6$ eV (see Fig.6.4, p.151). Separation of the two processes of fission and moderation
is needed, but the next obvious step of having all
the fuel at the centre surrounded by moderator is also
unsuitable. At thermal energies the effective cross-
section for removal of neutrons is
$2\cdot7 + (101 + 577)/140 \approx 7\cdot5$ b, weighting the cross-
sections according to the isotopic abundance, giving
an absorption length of order 2 cm. Thus the thermal-
neutron flux into the fissile centre is strongly
attenuated, and most of the fuel is inefficiently
used. The solution is therefore a matrix of natural
fuel rods a few millimetres in diameter embedded in
the moderator, through which must also be circulated
a coolant with low neutron-absorbing properties to
transfer the heat produced to the outside world. At
least this is the case for the graphite-moderated
reactors, the suitable coolant being carbon dioxide

gas; the liquid moderators can, in addition, be used
for heat transfer. Note that, because the fission
fragments have such a short range, virtually all the
heat is produced within the fuel itself, with a small
fraction arising from slowing down the neutrons and
absorption of γ-radiation.

The control of a thermal reactor derives once
again from the existence of delayed neutrons, though
the time constant when critical to prompt neutrons
is rather slower because the neutrons, when thermal-
ized diffuse rather slowly through the moderator
(thermal speeds are $\sim 10^5$ cm s^{-1} so diffusion speeds
are lower and distances of traverse to get back to
fuel are $\sim$ 10 cm giving a (prompt critical) rise
time $\sim$ ms). The mechanical control is in the form
of cadmium rods that can be inserted into the reactor.
Cadmium is a very strong absorber of thermal neutrons
by virtue of the large value of $\sigma(n,\gamma)$ ($\sim$ 20,000 b)
for the isotope $^{113}_{48}$Cd, of abundance 12·3 per cent.
This high cross-section arises from a resonance very
close to thermal energies, as has been discussed previously for ^{235}U.

So far only the rare isotope ^{235}U has been
utilized as a nuclear fuel. Obviously if ^{238}U could
also be used resources of power are considerably increased. This can be done, making use of the very
process that has proved troublesome in the design of
reactors discussed, namely, the capture process to
form ^{239}U. ^{239}U undergoes two quick β$^-$-decays to
^{239}Pu, which is long-lived ($\sim$ 24,000 years) and has
already been listed among nuclei which are thermally
fissile. This breeding of nuclear fuel can be best
achieved with a fast reactor since ν tends to rise

with neutron energy and the greater its value the
greater the surplus of neutrons for breeding over
and above the requirements for sustaining the reaction.
This surplus is made use of by making the core small
enough to allow a large leakage of neutrons into a
surrounding uranium blanket. Since the values of $\rho\nu$
are 2·45 and 2·70 for pure ^{235}U and ^{239}Pu, even after
reasonable loss incurred from other materials and
coolant, within the core, it is possible to breed
more new fuel than is consumed. It is necessary, of
course, to remove the uranium blanket from time to
time to separate out chemically the ^{239}Pu, which is
behaving effectively as a catalyst in the fission of
^{238}U. Thus, in principle, all of natural uranium
can be usefully consumed and at $\sim$ 200 MeV per nucleus
this works out at about 1g per day per MW!

FUSION IN THE SUN

Before discussing the nuclear physics behind
attempts to produce fusion in the laboratory, it is
of interest to describe briefly the nuclear reactions
going on inside the nearest working fusion reactor,
the Sun. Prior to the discovery of nuclear reactions
the Sun's source of energy was a mystery. The rate
of emission of radiant energy from the Sun is well
known and so also is its mass; gravitational collapse
of this mass provides a large source of energy, but
nowhere near large enough to keep the Sun radiating
for a few thousand million years, and there is geolo-
gical evidence on Earth to indicate that it has done
so for a considerable fraction of that time. On the
other hand, gravitational collapse to the present size
from a tenuous gas will provide enough energy to heat

the Sun up to a temperature $\sim 10^7$K even allowing for
radiation losses in the slow process. At that temperature, corresponding to a mean thermal energy
~ 1 keV, protons are beginning to collide sufficiently
violently with each other to penetrate the Coulomb
barrier. The stumbling block to the onset of nuclear
reactions culminating in the synthesis of helium from
hydrogen would appear to be the initial reaction of two
colliding protons. If we assume that two protons
momentarily make a compound nucleus, the diproton,
then this nucleus will decay back into the entrance
channel thereby producing no reaction. However, a
parallel though improbable decay is by β^+-emission to
the deuteron. Improbable though it may be, the Sun
contains a lot of colliding protons which cannot dis-
appear by any other process, so this one must occur
(though it has never been observed in the laboratory)
initiating the sequence:

$$p + p \rightarrow d + e^+ + \nu$$

$$p + d \rightarrow {}^3He + \gamma$$

$${}^3He + {}^3He \rightarrow {}^4He + 2p$$

Taking the first two reactions twice, the nett result
is

$$4p \rightarrow {}^4He + 2e^+ + 2\nu + Q, \qquad (7.1)$$

and from the mass values this liberates ~ 7 MeV per
proton absorbed (by comparison fission gives less than
1 MeV per nucleon involved). The choice of reactions
after the first is dictated by the preponderance of

protons. Thus the possible third reaction
d + ^{3}He → ^{4}He + p will not occur since all deuterons
formed will be mopped up by the second reaction before
they can get near a ^{3}He nucleus. On the other hand,
the ^{3}He + p interaction can only decay back into the
same channel, so ^{3}He is stable in the presence of .
protons and will build up in concentration until the
third reaction takes place at the appropriate rate
for equilibrium.

The rate at which a reaction proceeds depends
upon $n_1(v_1).n_2(v_2).\sigma(v_{12}).v_{12}$, where n_1 and n_2 are
velocity distributions of densities of particles 1
and 2 per unit energy range and v_{12} is their relative
velocity. The two Maxwellian distributions can be
combined into a single such function in terms of the
reduced mass and relative velocity giving (dropping
the suffixes) $n_1(v_1)n_2(v_2)v_{12} \propto v^3 \exp(-mv^2/2kT)$.
$\sigma(v)$ can be written as $\sigma_0.P(E).(E_B/E)$, where $P(E)$
is the Gamow factor, $\exp(-2\pi Z_1 Z_2 e^2/4\pi\epsilon_0 v)$, in the
penetrability for the limiting case of an energy E
well below the barrier height E_B, and therefore for a
nuclear radius which is very small compared to the
classical distance of closest approach. (E_B/E)
includes the $\pi\lambda^2$ term in the cross-section. Expressed
in this way σ_0 is the cross-section at $E \sim E_B$ and is
of the order of 0·1 b for a decay by particle emission
and perhaps $\sim$ 10 μb for decay by γ-emission.

The reaction rate is obtained by integrating over
all v. The two key factors will be the rapidly falling
$\exp(-E/E_T)$, where E_T, the thermal energy, is $\sim$ 1 keV
at 10^7 K, and the rapidly rising $\exp(-\beta E^{-\frac{1}{2}})$ where, for
$Z_1 = Z_2 = 1$ and E in keV, $\beta \sim 30$. Maximizing the
product of these two functions (the rate of variation

of other factors can be neglected) gives $E^{3/2} \sim \beta E_T/2$, giving $E \sim 6$ keV. Thus in the region around 6 keV there will be a sharp peak in the reaction rate (see Fig.7.2). At this energy the penetrability factor is $\sim \exp(-12)$ or 10^{-5} and the cross-section for p + d → ^{3}He + γ will be $\sim 10^{-10}$ b; the reaction

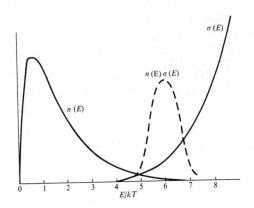

Fig.7.2. An indication of the peaking of the yield curve of a thermonuclear reaction.

^{3}He + ^{3}He → ^{4}He + 2p, having $Z_1 = Z_2 = 2$, will peak at ~ 15 keV and give a cross-section $\sim 10^{-10}$ b. For the p + p reaction the cross-section will in addition contain a factor representing the probability of β^+-decay during a nuclear characteristic time. A β^+-lifetime $\sim 10^3$s is a rough guess (about the same as the β^--decay of the neutron), and a nuclear time $\sim 10^{-21}$s gives the factor $\sim 10^{-24}$ and a cross-section at peak of about 10^{-29} b. Notice that none of these cross-sections is measurable in the laboratory; measurements could be pushed down to ~ 30 keV in

the second and $\sim$ 100 keV in the third reaction and
the energy-dependence used to extrapolate further
down. In the first reaction the factor 10^{-24} pre-
cludes measurement at any energy so the cross-section
must be derived theoretically.

An alternative set of reactions to the above
is the carbon-nitrogen cycle:

$$^{12}C + p \rightarrow {}^{13}N + \gamma \qquad ; \qquad {}^{13}N \rightarrow {}^{13}C + e^+ + \nu$$

$$^{13}C + p \rightarrow {}^{14}N + \gamma \qquad ; \qquad {}^{14}N + p \rightarrow {}^{15}O + \gamma \qquad (7.2)$$

$$^{15}O \rightarrow {}^{15}N + e^+ + \nu \qquad ; \qquad {}^{15}N + p \rightarrow {}^{12}C + {}^4He .$$

Thus the ^{12}C is regenerated and acts as a catalyst.
The importance of the two cycles depends upon the
temperature and upon the amount of ^{12}C present.
(Presumably the amount present in the primordial gas
from which the Sun condensed, since formation of
heavier elements requires a considerable concentration
of ^4He in order to start the process with
$3{}^4He \rightarrow {}^{12}C + \gamma$. This reaction is necessary since
$p + {}^4He$ merely returns to the same channel, as does
$^4He + He^4$.) At high enough temperatures (more massive
stars) this latter cycle will take over, since none
of the processes is particularly inhibited except
for the Gamow factor, which will be larger because
of the increased charges on C and N.

FUSION IN THE LABORATORY
Neither of the cycles described are suitable
bases for terrestrially controlled fusion. The hydrogen cycle begins with the improbable (weak) re-

action of two protons whilst the carbon-nitrogen cycle
requires higher temperatures which must be maintained
over periods long compared with the 10 min life-time
for β^+-decay of ^{13}N. The chief problems are the
heating up of the reacting materials, and, more
difficult, the containment of the hot material whilst
sufficient reaction occurs to make a nett liberation
of energy. A gas at 10^7 K (or strictly a plasma,
since it will be highly ionized) cannot simply be
retained by walls since, no matter what the material
of the container, it will not be able to withstand
bombardment of particles at such energies either
physically or chemically. Much work has been carried
out using 'magnetic bottles'; since the particles in
the plasma are charged they are deflected by magnetic
fields, and a great deal of cunning has gone into
designing the field so that the particles are kept
away from the walls of the vessel. In fact the magnetic fields can be used to feed energy into the
plasma and to compress it into a smaller and smaller
volume thereby increasing the reaction rate both by
temperature and density increase. Unfortunately, the
system becomes unstable, and so containment cannot
be maintained indefinitely. Containment times are
at present measured in fractions of a second, which
of course can be large enough to liberate enormous
amounts of energy as in the atom bomb or its later
development the hydrogen bomb, which is merely an
atom bomb surrounded by fusible material; the atom
bomb raises the temperature high enough ($\sim 10^9$ K)
and fusion then occurs more rapidly than the energy
and the material can be dissipated.

Obviously the reaction, or cycle of reactions,

must be intrinsically fast. A suitable one for the
laboratory would be

$$d + d \rightarrow {}^{3}\text{He} + n$$
$$\rightarrow T + p,$$

(7.3)

since it uses deuterium only which occurs as 1/7000 of
natural hydrogen and so is available in unlimited
supply. The reaction involves unit charges and will
therefore go at the lowest necessary temperatures
(10^{7} K in the Sun, but probably 10^{8} K for short containment times). These two reactions have rather
low Q-values (+ 3·27 MeV and + 4·03 MeV), releasing
only $\sim$ 1 MeV per nucleon. In addition, they have
relatively low peak cross-sections, $\sim$ 100 mb peaking
at several hundred keV. On the other hand, the
secondary reactions which could occur, namely
$d + {}^{3}\text{He} \rightarrow p + {}^{4}\text{He}$ and $d + T \rightarrow n + {}^{4}\text{He}$, have higher
Q-values (18·4 MeV and 17·58 MeV), and higher peak
cross-sections, especially the latter which peaks at
$\sim$ 80 keV with $\sigma \sim$ 7 b (see Fig.7.3). This latter
reaction therefore, would, go very quickly at 10^{8} K,
corresponding to a mean thermal energy of $\sim$ 10 keV,
at which energy the cross-section is still as high
as 10 mb. This mopping up, if complete, would boost
the energy output to $\sim$ 4 MeV per nucleon.

From the above, the most convenient reaction to
solve the problem of obtaining a useful source by
fusion is undoubtedly $d + T$ but it may be difficult
to use it as a basis for large-scale energy production. In principle, the tritium could be regenerated by the reaction

$$n + d \to T + Y \ (6 \cdot 26 \ \text{MeV})$$

or more conveniently by

$$n + {}^6\text{Li} \to T + {}^4\text{He} + 4 \cdot 78 \ \text{MeV}.$$

This latter reaction has a high thermal cross-section, so that ^{6}Li in the form of a solid compound could form a suitable regenerating blanket, as in some fusion bombs. There is no

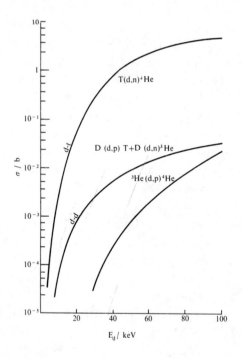

Fig.7.3. Thermonuclear reaction cross-sections. (Based on
          SEGRE (1964). *Nuclei and particles*, Benjamin, New York).

possibility, with these reactions, of breeding
more tritium than is consumed but, by addition
of Beryllium into the blanket, neutron
multiplication can result from the dominant
reaction

$$^9\text{Be} + n \rightarrow \alpha + \alpha + 2n - 1 \cdot 66 \text{ MeV},$$

leading to the possibility of tritium breeding.

A great deal depends upon the supply of materials
(assuming that the technical problems are solved).
The abundances in the earth's crust in parts per
million are, for some of the relevant elements:

H, $1400(^2\text{H}, 0 \cdot 2)$; Li, $65(^6\text{Li}, 4 \cdot 8)$; Be, 6;
and compare with U, 4 $(^{235}\text{U}, 0 \cdot 03)$.

These figures do not give a complete picture, since
hydrogen is readily available from water (the sea)
and deuterium separation is probably the simplest of
all isotope separations. Unless the breeding of
tritium is relatively simple, it would seem that
tritium-induced fusion will only be a stepping stone
towards the final solution based on the d + d reaction.

PROBLEMS

7.1. If neutrons (mass m) of kinetic energy E_0 are
    isotropically scattered by nuclei of mass M show
    that the scattered neutron energy-spectrum is
    independent of energy within the limits E_0 and
    $a^2 E_0$, where $a = (M-m)/(M+m)$. Show further that

the mean value of ln (E_0/E_1), where E_1 is the
energy after a single scatter, is equal to
$1 + (a \ln a)/(1 - a)$, denoted by ξ.

7.2. Extend the above result to the case where the
initial neutrons are not well defined in energy
but have a spectral distribution. Hence show
that, after n collisions, $\overline{\ln (E_0/E_n)} = n\xi$. If
the mean energy of the neutron fission spectrum
is 2 MeV, show that $\sim$ 18, 25, and 117 collisions
are required to thermalize neutrons in H, D,
and ^{12}C respectively.

7.3. An intimate mixture of ^{235}U and graphite is required for certain experiments. The graphite is
known to be contaminated with 1 p.p.m. by weight
of ^{10}B. What is the maximum fraction by weight
of ^{235}U in the mixture if the multiplication
factor at infinite size is not to exceed unity?
(σ_{abs} = 0·04 b, 3800 b, and 700 b respectively
for ^{12}C, ^{10}B, and ^{235}U at thermal energies; of
the last cross-section, σ_f accounts for 580 b.
Assume 2·5 neutrons per fission and that all
reactions take place at thermal energies only.)

7.4. For a gas consisting of two types of particles,
show that the distribution of *relative* velocities
of pairs consisting of one from each type is the
same as the absolute velocity distribution for a
single type of particle but with a mass equal
to the reduced mass, $m_1 m_2/(m_1 + m_2)$, where m_1,
m_2 are the masses of the two types.

7.5. (a)   If the fusion reaction
d + d → ^{3}He + n + 3·2 MeV takes place with
deuterons at rest, what is the kinetic energy of
the neutron?   (b) If the deuterons must come
within 10^{-11} cm of each other, what energy
must be supplied to overcome the electrostatic.
repulsion? (c) If this energy is supplied by
heating the deuterium to high temperature, what
order of magnitude of temperature is required?
(d) Discuss the effect of this heating on the
energies of the neutrons produced.

7.6. A nucleus consists of a large number N of particles
of spin $\frac{1}{2}$. Determine the number of ways, $R(M_S)$,
a given M_S can be produced assuming that $M_S \ll \frac{1}{2}N$
and that the Pauli Principle can be ignored.   By
identifying the difference $R(X) - R(X+1)$ with the
number of states of total spin $S = X$, show that
this number $\propto (2X+1)$. Hence the plausibility of
the statement made on p.173.

Appendix A: Cross-sections

To introduce the concept of cross-section, it is convenient to consider classically a nuclear reaction
in which a beam of particles falls on nuclei which
behave like completely absorbing spheres of cross-
sectional area σ. Assume, that one such nucleus is confined somewhere within an aperture of area A through
which passes a uniform beam of n particles per second.
The number absorbed per second will obviously be $n\sigma/A$.
If other nuclei are successively added to build up a
practical target of thickness t and nuclear number
density ρ (i.e. the number of nuclei per unit volume
of target material), then $\rho A t$ nuclei will be exposed
to the beam, and the number of incident particles
absorbed per second will be $\rho A t n\sigma/A = \sigma n\rho t$, provided
that $\sigma\rho t \ll 1$. (If this condition does not hold then
the beam flux is decreasing through the target and all
nuclei are not exposed to the same beam - under these
circumstances the beam decreases exponentially through
the target, in the form $n(t) = n(0) \exp(-\sigma\rho t)$). Notice
that the result does not depend upon the aperture area
A since, no matter what this area is, provided that
the target area is larger, each particle in the beam
passes through the same target thickness.

Thus, for the absorption process the number per
second absorbed is proportional to the cross-sectional
area of the nucleus. Replacing this completely absorbing nucleus by a hard sphere of the same size which
scatters all projectiles hitting it, the total number
per second scattered out of the beam will be the number

per second previously absorbed, $\sigma n \rho t$. The number per second scattered into solid angle $d\Omega$ at (θ, ϕ), in the usual nomenclature, relative to the beam will be $\sigma n \rho t f(\theta, \phi) d\Omega$, where $\int f(\theta, \phi) d\Omega = 1$, to ensure that the total scattering has the correct value. $f(\theta, \phi)$ is known as the angular distribution of the nuclear process. It is convenient to combine the product $\sigma f(\theta, \phi)$ into a single function $d\sigma(\theta, \phi)/d\Omega$, which is known as the differential cross-section for scattering along the direction (θ, ϕ). Most nuclear reactions are observed under conditions which are characterized by an axis of symmetry in the direction of the beam - so, in general, $\sigma(\theta, \phi)$ becomes $\sigma(\theta)$. Experiments may be carried out under conditions in which two axes are defined, e.g. the beam axis and an axis of polarization, and for these the generalized direction (θ, ϕ) is required.

These simple pictures can be extended to cover more complex situations: the nucleus may not be completely absorbing or scattering but partially the one, partially the other, and partially transparent; and, following absorption, something is usually emitted - the same, or another, type of particle or γ-radiation. These processes can all be included in a general definition of cross-section provided the equivalence of cross-section and nuclear cross-sectional area is revoked. This equivalence has arisen merely from the simple picture used to introduce the concept, and will in any case require modification when the wave nature of the projectile is taken into account. The fact remains, however, that the intrinsic nuclear property which determines the yield of a reaction has the dimensions of an area.

The general definition of cross-section for the reaction A + a → B + b is contained in the equation

$$\frac{dY_{a,b}(\theta)}{d\Omega} \, d\Omega = n\rho t \, \frac{d\sigma_{a,b}(\theta)}{d\Omega} \, d\Omega \quad , \qquad (A.1)$$

where $dY_{a,b}(\theta)/d\Omega$ is the number of particles of type b per second emitted into unit solid angle at an angle θ to the beam, and $d\sigma_{a,b}(\theta)/d\Omega$ is the differential cross-section for the process (a,b).

Also of use is the partial cross-section for the process (a,b) defined as

$$\sigma_{a,b} = \int \frac{d\sigma_{a,b}(\theta)}{d\Omega} \, d\Omega \quad .$$

The total cross-section is defined as $\sigma_{a,\text{tot}} = \sum_b \sigma_{a,b}$, where the summation is carried out over all possible reaction products B + b including the incident channel (known as elastic scattering). This latter cross-section is very useful when the projectiles are neutrons, but not so for charged particles, when it is dominated by the comparatively uninteresting but very large Coulomb cross-section.

Finally, notice that cross-section has been defined in terms of the flux of incoming particles that is their density times their velocity. In calculating reaction rates using perturbation theory the incoming particle wavefunction is normalized to represent one particle per unit volume, so in going to the cross-section a factor v is needed; σv is a measure of the reaction rate as calculated by perturbation theory. If the incoming particle is just above threshold, it may be possible that the appropriate matrix element varies only slowly with energy, as will also the density of final states if the reaction has positive Q-value. Under these circumstances $\sigma v \sim$ constant or $\sigma \propto 1/v$. This has been deduced in the text from the compound-nucleus formula, but as is seen above it has greater generality.

Appendix B: The need for a pairing term in the mass formula

In order to illustrate the need for a pairing term in the mass formula, it is convenient to neglect all terms in the mass equation but the Coulomb and symmetry terms and to simplify these to the form

$$a(A-2Z)^2 + bZ(Z-1) \qquad (a, b \text{ both positive}) \quad .$$

The conditions that (A,Z) is stable relative to its neighbours $(A, Z-1)$ and $(A, Z+1)$ is

$$a(A-2Z)^2 + bZ(Z-1) < a(A-2Z-2)^2 + b(Z+1)Z$$

and also

$$< a(A-2Z+2)^2 + b(Z-1)(Z-2) \quad .$$

These can be re-expressed as

$$b.2Z > 4a\{A-2Z-1\}$$

and

$$b.2Z < 4a\{(A-2Z-1) + (1 + \frac{N}{Z-1})\} \quad .$$

Adding a nucleon to (A,Z) gives:

adding p, the energy change $E_p = -a(2A-4Z-1) + b.2Z$,

adding n, the energy change $E_n = +a(2A-4Z+1)$.

The condition for $(A+1,Z)$ to be stable relative to

$(A+1,Z+1)$ is that $E_p > E_n$, which simplifies to

$$b.2Z > 4a\{(A-2Z-1) + 1\} .$$

For the second neutron to produce a stable nucleus, the condition is that

$$M(A+2,Z) < M(A+2,Z+1),$$

which is equivalent to increasing A to $A+1$ in the last inequality, i.e.

$$b.2Z > 4a\{(A-2Z-1)\}+ 2 .$$

These conditions may be expressed in terms of an allowable variation of the function $b.2Z/4a$ as:

for (A,Z) stable, the range is $0 \rightarrow 1 + \dfrac{N}{Z-1}$ from the value $(A-2Z-1)$,

for first neutron, the range is $1 \rightarrow 1 + \dfrac{N}{Z-1}$ from the value $(A-2Z-1)$,

for second neutron, the range is $2 \rightarrow 1 + \dfrac{N}{Z-1}$ from the value $(A-2Z-1)$.

Since N/Z varies from 1 for light nuclei to $\sim 1\frac{1}{2}$ for the heaviest nuclei, these ranges are, for the heavy nuclei $2\frac{1}{2}$ units, $1\frac{1}{2}$ units, and $\frac{1}{2}$ unit respectively. Thus the probability of a neutron adding is roughly $\frac{3}{5}$, and a further neutron roughly $\frac{1}{3}$. A similar simple argument with protons would lead to $\frac{2}{5}$ for the first proton and zero for the second. Using all the terms of the mass equation and the more exact functions for the two discussed does not alter the general con-

clusion that if (A,Z) is stable then the addition
of two more nucleons is more likely to be of the form
(p,n) for stability than (2p) or (2n). Thus odd-odd
nuclei should be as prolific as even-even.

Obviously a term is missing from the mass equation. It must be such that if A and Z are both even
then, whilst not altering the probability that the
first particle be a neutron for stability ($\frac{3}{5}$ is just
about correct for a heavy nucleus), it must increase
the probability for a second neutron giving stability
up to unity. This extra term is not needed in considering the stability of odd-A nuclei, but must be
chosen such that in going from even A to $A+2$, where
both nuclei are stable, pairs of like particles will
be added. Hence the form adopted.

Appendix C: Exchange forces

To introduce this concept it is interesting to consider the single ionized H_2 molecule - a system of two protons and one electron. When the separation

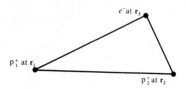

Fig.C.1. Schematic representation of the hydrogen-molecule ion.

$|\underline{R}| = |\underline{r}_2 - \underline{r}_1|$ is very large the electron wavefunction will look like the hydrogen-atom bound state situated at $\underline{r}_1$ or $\underline{r}_2$ and designated by $\phi(\underline{r}_3-\underline{r}_1)$ or $\phi(\underline{r}_3-\underline{r}_2)$. However, the correct wavefunction must be symmetric (or antisymmetric) with respect to the two protons. Therefore, put $\psi = \alpha_{\pm}^{-\frac{1}{2}} \{\phi(\underline{r}_3-\underline{r}_1) \pm \phi(\underline{r}_3-\underline{r}_2)\}$, where the normalization integral α takes the value 2 as $R \rightarrow \infty$, whilst $\alpha_+ \rightarrow 4$, $\alpha_- \rightarrow 0$ as $R \rightarrow 0$. The energy integral is

$$4\pi\epsilon_0 \frac{V}{e^2} - \frac{1}{R} = -\int \psi^* \left(\frac{1}{|\underline{r}_3-\underline{r}_1|} + \frac{1}{|\underline{r}_3-\underline{r}_2|} \right) \psi d\tau_3 =$$

$$= -2 \int \frac{\psi^* \psi}{|\underline{r}_3-\underline{r}_1|} d\tau_3 ,$$

by symmetry. Substituting for ϕ gives four integrals, and for ϕ real (corresponding to a bound state) two are the same giving (where $\underline{r} = \underline{r}_3-\underline{r}_1$)

$$4\pi\varepsilon_0 \frac{V}{e^2} - \frac{1}{R} = -2\alpha_{\pm}^{-1}\{\int\phi^2(\underline{r})\frac{1}{r}\,d\tau + \int\phi^2(\underline{r})\frac{1}{|\underline{r}-\underline{R}|}\,d\tau \pm$$

$$\pm\ 2\int\phi(\underline{r}-\underline{R})\phi(\underline{r})\frac{1}{r}\,d\tau\}. \quad (C.1)$$

The first term represents the binding energy of a hydrogen atom, the second term represents the electronic contribution to the Coulomb energy of a proton distant R from a hydrogen atom; for large R this term exactly cancels the proton repulsion ($1/R$ on the left-hand-side), but for small R it becomes the finite binding energy of the hydrogen atom. The last term, unlike the other two, has no classical analogue; it effectively arises from not knowing to which proton the electron belongs. Its sign depends upon the overlap of the two atomic wavefunctions when separated by R and, if the wavefunctions have nodes, can be positive or negative. For the lowest s-state there are no nodes, so the integral is always positive, giving increased binding for the symmetric wavefunction and reduced for the antisymmetric. At large distances bound wavefunctions fall off exponentially. If the s-state looks like $\exp(-\kappa r)$ at large r, the main contribution to the integral will occur nearly halfway between the two protons and will obviously contain $\exp(-\kappa R)$ in the product of the wavefunctions (this qualitative argument neglects the effects of terms like $1/r$ in the integral). Thus the integral will also behave like $\exp(-\kappa R)$ at large distances.

Now assume that the existence of the electron is not known, but that experiments can be performed to determine the potential between a proton (charged) and a hydrogen atom (neutral) as a function of separation R. The classical term will be constant at large R (but

unknown since the structure of hydrogen is unknown);
however, at small R, the proton gets inside the electron cloud and experiences a net repulsion by the
other proton. For the symmetric case, the non-
classical term increases exponentially (approximately)
to a finite binding at $R = 0$. The state of affairs is
presented graphically in Fig.C.2.

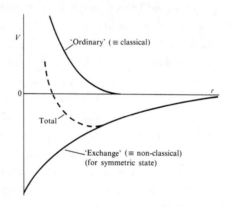

Fig.C.2. Potential of hydrogen-molecule ion. $V=0$ corresponds
        to infinite separation between ionized atom (p) and
        neutral atom (H).

    Thus there eixsts a potential between the proton
and the hydrogen atom which depends upon the spatial
symmetry of the (p,H) system. During the interaction
the proton and the hydrogen atom cannot be individually
identified, and this is interpreted as exchange of
charge and spin (in the form of an unknown electron).
From Fig.C.2 this can result in attraction at large
R, whilst the repulsion at small R arises because the
'fundamental-particle' hydrogen atom in fact has a
structure.

The analogue with nucleons is obvious, at least
for the (p,n) system. A similar argument to the above
can be applied if n is looked upon as p + π^- and gives
the correct range for the nuclear force. Since the
proton and neutron have almost the same mass, the total
energy (rest + binding) of the bound π^- must be nearly
zero. But $\kappa = |p|/\hbar$ in the unphysical region, where
$|p|$ is given by $p^2c^2 = E^2 - m_\pi^2 c^4$ with $E = 0$. Therefore
$\kappa = m_\pi c/\hbar = (1 \cdot 4 \text{ fm})^{-1}$. In order to produce charge-
independence, a π^0 must be introduced (to give the
(p,p) and (n,n) interactions) and also a π^+ (though
its need is not immediately obvious; it arises because
the above picture has made the proton look 'more fundamental' than the neutron - the π^+ redresses the
balance since its exchange appears to make the proton
look like n + π^+).

Having gone through the above reasoning, it is
convenient to forget the mechanism and express the
different exchanges mathematically. The operator
$\underline{s}_1 \cdot \underline{s}_2 \equiv \frac{1}{2}\{(\underline{s}_1 + \underline{s}_2)^2 - \underline{s}_1^2 - \underline{s}_2^2\} = \frac{1}{2}\{\underline{S}^2 - \underline{s}_1^2 - \underline{s}_2^2\}$ takes the value
$\frac{1}{2}\{S(S+1) - s_1(s_1+1) - s_2(s_2+1)\}$ for the spin state S, re-
sulting from the combination of two spin states s_1 and
s_2. For $s_1 = s_2 = \frac{1}{2}$, it takes the value $+\frac{1}{4}$ for $S = 1$
and $-\frac{3}{4}$ for $S = 0$. Thus the operator
$P^\sigma \equiv \frac{1}{2} + 2\underline{s}_1 \cdot \underline{s}_2 = 1$ for $S = 1$, triplet state, and
$= -1$ for $S = 0$, singlet state. This P^σ is the spin-
exchange operator. If $\uparrow\downarrow$ is used to denote the state
in which particle 1 has spin up, particle 2 spin down,
then it can be represented thus:

$$\uparrow\downarrow = \frac{1}{2}(\uparrow\downarrow + \downarrow\uparrow) + \frac{1}{2}(\uparrow\downarrow - \downarrow\uparrow) = \frac{1}{\sqrt{2}}\{(S=1) + (S=0)\},$$

therefore

$P^{\sigma}\uparrow\downarrow = \frac{1}{\sqrt{2}}\{(S=1) - (S=0)\} = \frac{1}{2}(\uparrow\downarrow+\downarrow\uparrow) - \frac{1}{2}(\uparrow\downarrow-\downarrow\uparrow) = \downarrow\uparrow .$

Thus P^{σ} has exchanged spins. P^{T}, which exchanges charge, can be defined similarly. The general potential is then (see Chapter 2, p. 40, for nomenclature) $V = V_W + V_H \, P^T + V_B \, P^{\sigma} + V_M \, P$ (where $P = -P^T P^{\sigma}$ is a requirement of the Pauli principle). Since the exchange operator has a purely numerical part, the potential can be rewritten

$$V = V_1 + V_2 \, \underline{t}_1 \cdot \underline{t}_2 + V_3 \underline{s}_1 \cdot \underline{s}_2 + V_4 (\underline{s}_1 \cdot \underline{s}_2)(\underline{t}_1 \cdot \underline{t}_2)$$

and the operators are (loosely) referred to as exchange operators.

The effect of the operators, P^{σ}, P^{T}, P^{S}, is shown in Table C.1 for S- and P- states. Notice that it is the pattern of the changes of sign which is important. The sign of a whole column can be changed merely by changing the sign of the potential function accompanying the operator.

TABLE C.1

ψ	Symmetry[††]			Sign of operator[†]			
	Spin	Isospin	Space	W	H	B	M
$^3S_1$	s	a	s	+	−	+	+
$^1S_0$	a	s	s	+	+	−	+
$^3P_{012}$	s	s	a	+	+	+	−
$^1P_1$	a	a	a	+	−	−	−

[†] see p.40 for definition of W,H,B,M.
[††] s denotes 'symmetric'; a denotes 'antisymmetric'.

Appendix D: Spin matrices

Angular momentum can be introduced into quantum mechanics by inserting the operator form $\underline{p} = -i\hbar\nabla$ into the classical expression $\underline{J} = \underline{r}\wedge\underline{p}$, to give $\underline{J} = -i\hbar(\underline{r}\wedge\nabla)$. It is left to the reader to show that $\underline{J}\wedge\underline{J} = i\hbar\underline{J}$ (hint: use Cartesian coordinates), i.e. $M_x M_y - M_y M_x = i\hbar M_z$ and two others, cyclically.

Although these equations have been derived from considerations of orbital angular momentum which is quantized to integer values, Dirac has used them to show that odd half-integer values are also permitted indicating the existence of another type of intrinsic angular momentum. Here it is proposed to consider $J = \frac{1}{2}$, and to designate it s, the intrinsic spin angular momentum of a fundamental particle. For this particular case it can be shown that $s_x s_y + s_y s_x = 0$, etc. and therefore $s_x s_y = (i\hbar/2)s_z$, etc. Since s_z is two-valued, having values $\pm\frac{1}{2}\hbar$, the wavefunction representing an arbitrary state of spin can be written as a two-component vector $\begin{bmatrix} a \\ b \end{bmatrix}$ in an appropriate two-dimensional space. Ths components of $\underline{s}$, namely, s_x, s_y, s_z will be represented by 2×2 matrices, of which s_z will be diagonal and of the form $\frac{1}{2}\hbar\begin{bmatrix} 1 & 0 \\ 0 & -1 \end{bmatrix}$ to give the required eigenvalues when acting on the eigenfunctions $\begin{bmatrix} 1 \\ 0 \end{bmatrix}$ and $\begin{bmatrix} 0 \\ 1 \end{bmatrix}$ representing 'spin up' and 'spin down'. Putting $s_x \equiv \begin{bmatrix} a & b \\ c & d \end{bmatrix}$, $s_y \equiv \begin{bmatrix} e & f \\ g & h \end{bmatrix}$, and using $s_y s_z = \frac{1}{2}i\hbar s_x$ and $s_z s_x = \frac{1}{2}i\hbar s_y$ results in

$$a = e = d = h = 0, \quad g = ic, \quad f = -ib.$$

There is some latitude in the choice of s_x, s_y; the forms commonly chosen are

$$s_x = \tfrac{1}{2}\hbar \begin{bmatrix} 0 & 1 \\ 1 & 0 \end{bmatrix} \text{ and } s_y = \tfrac{1}{2}\hbar \begin{bmatrix} 0 & -i \\ i & 0 \end{bmatrix} .$$

Notice that $s_x^2 + s_y^2 + s_z^2 = s^2 = \tfrac{3}{4}\hbar^2$, which conforms to the usual expression $J^2 = \hbar^2 J(J+1)$.

The raising operator

$$\hbar^{-1} s^+ \equiv \hbar^{-1}(s_x + i s_y) = \begin{bmatrix} 0 & 1 \\ 0 & 0 \end{bmatrix}$$

obviously operates on $\begin{bmatrix} 1 \\ 0 \end{bmatrix}$ to give $\begin{bmatrix} 0 \\ 0 \end{bmatrix}$ and on $\begin{bmatrix} 0 \\ 1 \end{bmatrix}$ to give $\begin{bmatrix} 1 \\ 0 \end{bmatrix}$; hence its name. Similarly the lowering operator $\hbar^{-1} s^-$ is defined as $\hbar^{-1}(s_x - i s_y)$.

In a similar way the components of isospin may be defined. As for spin, the two-component isospin space has no connection with any classical space, but, and in contrast to spin, neither does the three-component isospin space in which the components τ_x, τ_y, τ_z are defined. There is a one-to-one correspondence between the z-component of isospin and the charge of the nucleon, but otherwise there is no further definition of this space. The isospin raising and lowering operators have the property of changing one nucleon type to its counterpart, without changing any other part of its wavefunction. It should be stated that this is a rather restricted view of isospin since only nucleons are under consideration as distinct from its greater ramifications when dealing with other hadrons.

Bibliography

Other textbooks recommended for further reading,
some of which have been referred to in the text, are:

*Blatt, J.M. and Weisskopf, V. (1952). *Theoretical
nuclear physics.* Wiley, New York.

Bowler, M.G. (1973). *Nuclear physics.* Pergamon Press,
Oxford.

Burcham, W.E. (1963). *Nuclear physics, An introduction.*
Longman, London.

*Dirac, P.A.M. (1958). *Principles of quantum mechanics.*
Clarendon Press, Oxford.

Enge, H. (1966). *Introduction to nuclear physics.*
Addison-Wesley, New York.

Evans, R.D. (1955). *The atomic nucleus.* McGraw-Hill,
New York.

Fermi, E. (1950). *Nuclear physics.* Chicago University
Press.

*Hodgson, P.E. (1963). *The optical model of elastic
scattering.* Clarendon Press, Oxford.

Kuhn, H.G. (1970). *Atomic spectra.* Longmans, London.

Paul, E.B. (1969). *Nuclear particle physics.* North-
Holland, Amsterdam.

Pauling, L. and Wilson, E.B. (1935). *Introduction to
quantum mechanics.* McGraw-Hill, New York.

Perkins, D.H. (1972). *Introduction to high-energy
physics.* Addison-Wesley, New York.

*Preston, M.A. (1962). *Physics of the nucleus.*
Addison-Wesley, New York.

*Schiff, L.E. (1968). *Quantum mechanics*. McGraw-Hill,
  New York.

Segre, E. (1964). *Nuclei and particles*. Benjamin,
  New York.

Those indicated by an * are postgraduate in standard.

Answers to problems

1.1. $n \sim 3$.

1.4. $\varepsilon \sim 0 \cdot 63$, $\gamma \sim 19 \cdot 1$.

1.5. $3 \cdot 7$ fm.

2.2. $+ 0 \cdot 31$ nm.

2.5. $\sim 2 \cdot 07$ MeV.

3.4. $I \sim \frac{2}{3} I_{rigid}$.

3.6. $\lambda \sim 16$, $4 \cdot 8$, $6 \cdot 3$, 16 fm respectively.

4.4. ~ 3 s.

5.1. $\sim 6 \times 10^{-18}$ s.

6.2. $7 \cdot 0191$ a.m.u.

6.6. $8 \cdot 5$ µb.

7.3. $0 \cdot 00115$.

7.5. (a) $2 \cdot 4$ MeV; (b) 14 keV in Centre-of-mass system;
(c) $\sim 1 \cdot 6 \times 10^{8}$ K; (d) thermal spread $\sim \pm 100$ keV.

Index

Physical constants and conversion factors

Avogadro constant	L or N_A	$6 \cdot 022 \times 10^{23}$ mol^{-1}
Bohr magneton	μ_B	$9 \cdot 274 \times 10^{-24}$ J T^{-1}
Bohr radius	a_0	$5 \cdot 292 \times 10^{-11}$ m
Boltzmann constant	k	$1 \cdot 381 \times 10^{-23}$ J K^{-1}
charge of an electron	e	$-1 \cdot 602 \times 10^{-19}$ C
Compton wavelength of electron	$\lambda_C = h/m_e c =$	$2 \cdot 426 \times 10^{-12}$ m
Faraday constant	F	$9 \cdot 649 \times 10^4$ C mol^{-1}
fine structure constant	$\alpha = \mu_0 e^2 c/2h =$	$7 \cdot 297 \times 10^{-3}$ ($\alpha^{-1} = 137 \cdot 0$)
gas constant	R	$8 \cdot 314$ J K^{-1} mol^{-1}
gravitational constant	G	$6 \cdot 673 \times 10^{-11}$ N m^2 kg^{-2}
nuclear magneton	μ_N	$5 \cdot 051 \times 10^{-27}$ J T^{-1}
permeability of a vacuum	μ_0	$4\pi \times 10^{-7}$ H m^{-1} exactly
permittivity of a vacuum	ϵ_0	$8 \cdot 854 \times 10^{-12}$ F m^{-1} ($1/4\pi\epsilon_0 =$ $8 \cdot 988 \times 10^9$ m F^{-1})
Planck constant	h	$6 \cdot 626 \times 10^{-34}$ J s
(Planck constant)/2π	$\hbar$	$1 \cdot 055 \times 10^{-34}$ J s $= 6 \cdot 582 \times 10^{-16}$ eV s
rest mass of electron	m_e	$9 \cdot 110 \times 10^{-31}$ kg $= 0 \cdot 511$ MeV/c^2
rest mass of proton	m_p	$1 \cdot 673 \times 10^{-27}$ kg $= 938 \cdot 3$ MeV/c^2
Rydberg constant	$R_\infty = \mu_0^2 m_e e^4 c^3/8h^3 =$	$1 \cdot 097 \times 10^7$ m^{-1}
speed of light in a vacuum	c	$2 \cdot 998 \times 10^8$ m s^{-1}
Stefan–Boltzmann constant	$\sigma = 2\pi^5 k^4/15h^3 c^2 =$	$5 \cdot 670 \times 10^{-8}$ W m^{-2} K^{-4}
unified atomic mass unit (^{12}C)	u	$1 \cdot 661 \times 10^{-27}$ kg $= 931 \cdot 5$ MeV/c^2
wavelength of a 1 eV photon		$1 \cdot 243 \times 10^{-6}$ m

$1 \text{ Å} = 10^{-10}$ m; $\quad 1$ dyne $= 10^{-5}$ N; $\quad 1$ gauss (G) $= 10^{-4}$ tesla (T);
$0°$C $= 273 \cdot 15$ K; $\quad 1$ curie (Ci) $= 3 \cdot 7 \times 10^{10}$ s^{-1};
1 J $= 10^7$ erg $= 6 \cdot 241 \times 10^{18}$ eV; $\quad 1$ eV $= 1 \cdot 602 \times 10^{-19}$ J; $\quad 1$ cal$_{th} = 4 \cdot 184$ J;
$\ln 10 = 2 \cdot 303$; $\quad \ln x = 2 \cdot 303 \log x$; $\quad$ e $= 2 \cdot 718$; $\quad \log$ e $= 0 \cdot 4343$; $\quad \pi = 3 \cdot 142$